Current Topics in Pathology
92

Springer

Berlin
Heidelberg
New York
Barcelona
Hong Kong
London
Milan
Paris
Singapore
Tokyo

C. L. Berry (Ed.)

Transplantation Pathology

A Guide for Practicing Pathologists

Contributors
E. Alvarez-Fernández, K. Atkinson, S. M. Dodd, D. A. Hilton
C. C. Kibbler, G. Koukoulis, S. Love, J. P. Mindán, A. Panizo
B. Portmann, E. Thorsby, F. Vartdal

Springer

Dr. med. Beat Thürlimann
Kantonsspital St. Gallen
Klinik C für Innere Medizin
Abteilung für Onkologie und Hämatologie
CH-9007 St. Gallen

ISBN-13:978-3-642-64162-6 e-ISBN-13:978-3-642-59877-7
DOI: 10.1007/978-3-642-59877-7

ISSN 0080-0015

Library of Congress Cataloging-in-Publication Data
Bisphosphonates in clinical oncology: the development in pamidronate/B. Thürlimann (ed.). p. cm. – (Recent results in cancer research, ISSN 0080-0015; 149). Includes bibliographical references and index.ISBN-13:978-3-642-64162-6(hardcover: alk.paper).1.Disodium pamidronate. 2. Bones–Cancer–Chemotherapy. 3. Diphosphonates–Therapeutic use. I. Thürlimann, B. (Beat), 1955–. II. Series. [DNLM: 1. Diphosphonates-pharmacology. 2. Antineoplastic Agents–pharmacology. 3. Hypercalcemia–drug therapy. 4. Bone Neoplasms–drug therapy. 5. Bone Neoplasms–secondary. 6. Bone and Bones–physiology. [W1 RE106P v. 149 1998/QZ 267 B6157 1998] RC261R35 vol. 149 [RC280.B6] 616.99′4 s [616.99′4061]–DC21 DNLM/DLC for Library of Congress.

Softcover reprint of the hardcover 1st edition 1999

Production: PRO EDIT GmbH, D-69126 Heidelberg
Typesetting: K+V Fotosatz GmbH, D-64743 Beerfelden

SPIN 10643981 19/3133-5 4 3 2 1 0 – Printed on acid-free paper

List of Contributors

ALVAREZ-FERNÁNDEZ, E. MD, PhD	Professor Titular de Anatomía Patológica, Head Departamento de Anatomía Patológica Hospital General Universitario Gregorio Marañon Doctor Esquerdo, 46 28007 Madrid, Spain
ATKINSON, K. MD, FRCP, FRACP	Director of Clinical Transplantation SyStemix, Inc. 3155 Porter Dr. Palo Alto, CA 94304, USA
DODD, S. M. Dr.	Senior Lecturer in Department of Morbid Anatomy and Histopathology St. Bartholomew's and the Royal London School of Medicine and Dentistry Royal London Hospital Whitechapel London E1 1BB, UK
HILTON, D. A. Prof.	Consultant Neuropathologist Department of Histopathology Derriford Hospital Plymouth PL6 8DH, UK
KIBBLER, C. C. MA, FRCP, FRCPath	Consultant in Medical Microbiology Royal Free Hospital Pond Street London NW3 2QG, UK
KOUKOULIS, G.	*Formerly:* Institute of Liver Studies King's College Hospital Denmark Hill London SE5 9 RS, UK

LOVE, S.
Prof.

Professor of Neuropathology
Department of Neuropathology
Frenchay Hospital
Bristol BS16 1LE, UK

MINDÁN, J.P.
Prof.

Departamento de Anatomía Patológica
Clínica Universitaria
Universidad de Navarra
31080 Pamplona, Spain

PANIZO, A.
Prof.

Departamento de Anatomía Patológica
Clínica Universitaria
Universidad de Navarra
31080 Pamplona, Spain

PORTMANN, B.
MD, FRCPath

Professor of Hepatopathology and Hon.
Consultant Histopathologist
Institute of Liver Studies
King's College Hospital
Denmark Hill
London SE5 9 RS, UK

THORSBY, E.
Prof.

Institute of Transplantation Immunology
The National Hospital
and University of Oslo
Rikshospitalet
0027 Oslo, Norway

VARTDAL, F.
Prof.

Institute of Transplantation Immunology
The National Hospital
and University of Oslo
Rikshospitalet
0027 Oslo, Norway

Preface

The widespread use of transplantation in an increasing range of diseases has meant that many pathologists, other than those in special centres, are seeing material from transplanted patients. It is also increasingly necessary for those in training to understand the essential features of transplant related pathology; many of the institutions in which they will work will have transplant programmes.

This volume attempts to give a clear account of how patients with the commonly transplanted tissues and organs are managed by the use of pathological techniques. The emphasis is on the important complications of rejection and infection and in order to understand these, an outline of what is of basic scientific importance in rejection is given. Finally, in a section on the central nervous system, the way in which transplantation may be used to inform us in terms of development, the pathogenesis of disease and with new developments in a group of conditions resistant to therapy is illustrated.

The criteria for the evaluation of rejection are given, in terms of recent internationally agreed criteria, and there has been a deliberate attempt to give a critical valuation of the likelihood of the various infectious complications. It is hoped that this volume will be a useful reference source for the specialist and that it will inform the generalist.

London, November 1998

SIR COLIN BERRY

Contents

Transplantation Immunology – The Role of Human Leucocyte Antigen in Allorecognition

F. Vartdal and E. Thorsby

1 Introduction

Transplantation of organs and bone marrow is the therapy of choice for an increasing number of patients. The greatest problem encountered after transplantation of cells and tissue from a genetically different donor is that the immune system of the recipient recognizes the transplant as foreign and will try get rid of it in essentially the same way as the immune system deals with invading organisms. The result is a host-versus-graft (HVG) response, which may lead to rejection of the transplant. Similarly, T lymphocytes (T cells for short) transferred along with the stem cells in a donor bone marrow graft can recognize host cells as foreign and will initiate a graft-versus-host (GVH) response, which may cause serious and often fatal GVH disease.

Human leucocyte antigen (HLA) molecules play an instrumental role in triggering both HVG and GVH immune responses. In this chapter, we first describe the

Current Topics in Pathology
Volume 92, C. L. Berry (Ed.)

structure and normal biological function of HLA molecules and how these molecules are involved in the recognition of foreign cells, i.e., in allorecognition. The activation and effector mechanisms of alloimmune responses are then treated briefly, followed by a discussion of the role of HLA matching and of the various strategies used to induce donor-specific tolerance.

2 Types of Transplantation

There are several types of transplantation and in each of these the transplanted tissue is perceived differently by the immune system:

- Autotransplantation. Cells or tissue are transplanted within the same individual. One example is autotransplantation of skin in the treatment of burns; another is harvesting bone marrow from patients before they are given sublethal doses of cytostatic drugs and/or irradiation, followed by re-infusion of the patients' own bone marrow.
- Isotransplantation: occurs between monozygotic (MZ) twins. In MZ twins, all genes are identical [except for rearranged and somatic mutated V, D and J genes of the complexes encoding the immunoglobulin (Ig) and T cell receptors (TCR)].
- Allotransplantation: occurs between genetically different individuals within the same species.
- Xenotransplantation: occurs between individuals belonging to different species.

Following autotransplantation and isotransplantation, no HVG or GVH responses are seen. Thus, immunosuppressive treatment is not necessary. Following allo- and xenotransplantation, strong HVG and GVH responses are seen, necessitating immunosuppressive treatment.

3 Types of Rejection

Acute rejection is caused by primary activation of alloreactive T cells and B lymphocytes towards donor histocompatibility molecules or fragments of them. It is usually fully developed 5–7 days after transplantation, if the recipient is not treated with immunosuppressive drugs.

Accelerated acute rejection is caused by reactivation of T and B cells that have been activated previously by cells bearing the same (or similar) histocompatibility molecules as those present in the actual graft. It may develop a few days after transplantation.

Hyperacute rejection is caused by preformed alloreactive antibodies to blood group (ABO) or histocompatibility (HLA) antigens present on donor endothelial cells, or by preformed xenoreactive antibodies that are reactive with endothelial cell antigens. It may develop from minutes to a few hours after transplantation.

Chronic rejection is caused by alloreactive T and B cells as well as by alloantibodies, and may develop in patients who have experienced several prior acute rejection episodes. It may develop from months to years after transplantation.

4 Structure and Function of Human Leucocyte Antigen (HLA) Molecules

Whereas B cells recognize foreign molecules on the surface of cells and in the extracellular space directly, the function of T cells is to survey intracellular proteins that are either synthesized within the cytosol or taken up by endocytosis (see Fig. 1). The function of major histocompatibility complex (MHC) molecules (called HLA in humans) is to bind fragments of intracellular proteins and then transport them to the cell surface, where they are displayed to T cells. Thus, HLA molecules act as informers for T cells.

There are two main classes of HLA molecule: HLA class I and class II. Both are heterodimeric molecules formed by an α- and a β polypeptide chain. With the exception of the HLA class-I β chain (β2-microglobulin), which is encoded by a gene on chromosome 15, HLA molecules are encoded by genes of the HLA complex, which is located on the short arm of chromosome 6 (see Fig. 2). In addition to those genes that encode the HLA molecules, the HLA complex encompasses many genes that encode molecules with other immunological functions. For example, there are tumour necrosis factor (TNF) genes, some complement genes, as well as molecules that are involved in the processing of proteins/peptides to be bound by HLA-molecules, such as large multifunctional proteasome (LMP), transporters associated with antigen processing (TAP) and HLA-DM molecules (see Fig. 2) (reviewed in [14]). The HLA complexes of each of the two chromosomes in a pair is called an HLA haplotype. Thus, each individual has two HLA haplotypes, one inherited from each parent.

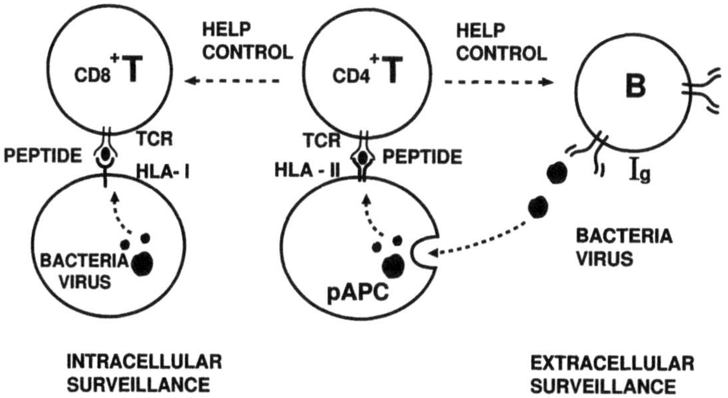

Fig. 1. A schematic illustration of specific immune responses, in which T cells (*T*) survey the intracellular milieu and the B cells (*B*) survey the extracellular milieu. $CD4^+$ T cells (helper cells) are able to recognize peptides from endocytosed proteins when these peptides are bound and displayed to the $CD4^+$ T cell by the human leucocyte antigen (HLA) class-II molecules of a professional antigen-presenting cell (*pAPC*). $CD8^+$ T cells (cytotoxic cells) recognize peptides derived from cytosolic proteins when these peptides are displayed by HLA class-I molecules of cells. B cells are able to react directly with antigenic determinants (epitopes) on intact foreign extracellular material (free or membrane-bound). $CD4^+$ T cells help and control activation of $CD8^+$ T cells and B cells

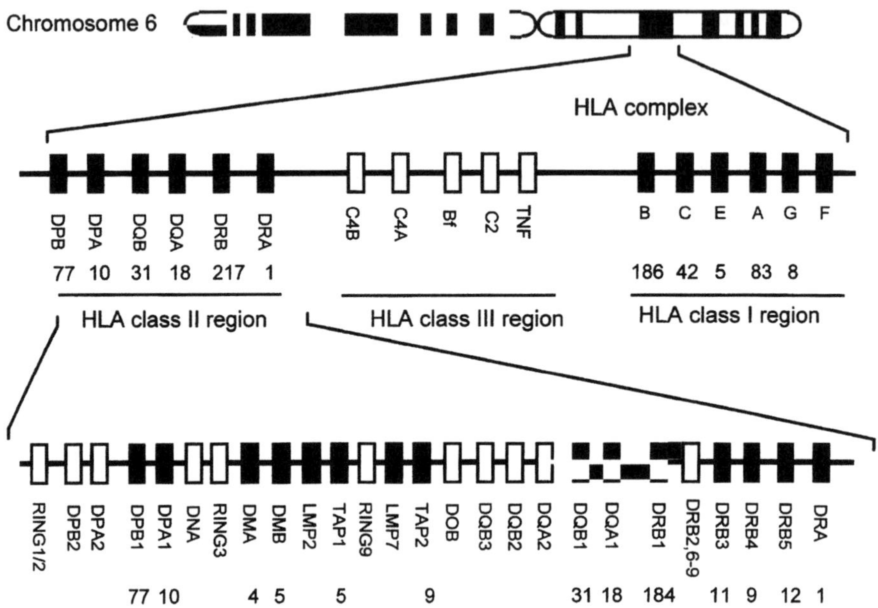

Fig. 2. A simplified map of the human leucocyte antigen (*HLA*) gene complex, showing the HLA class-I, class-II and class-III regions. The HLA class-II region is also shown in greater detail at the *bottom*. Genes encoding HLA molecules and molecules involved in antigen processing (TAP1, TAP2, LMP2, LMP7, DMA and DMB) are depicted as *black boxes*, whereas other genes (complement genes in the class-III region, pseudogenes and genes with undefined functions) are depicted as *open boxes*. The *numbers* of hitherto defined alleles of HLA genes and processing genes are also given (according to Bodmer et al. [10])

The function of the classical HLA class-I molecules, which include HLA-A, -B and -C, is to present peptides to $CD8^+$ T cells (cytotoxic cells), whereas the function of HLA class-II molecules, HLA-DR, -DQ and -DP, is to present peptides to $CD4^+$ T cells (helper T cells) [59]. HLA class-I molecules bind and present peptides generated from proteins that have been synthesized in the cytosol (reviewed in [36]). These proteins are degraded into peptides by LMP, after which the resulting peptides are transported by TAP molecules into the endoplasmic reticulum (ER). Here, the peptides are bound to class-I molecules and are subsequently transported to the cell surface. HLA class-II molecules bind and present peptides generated either from proteins that have been taken up from the extracellular fluid or from cell surface proteins taken up by the cell by endocytosis (reviewed in [13]). As HLA class-II molecules are synthesized in the ER, they bind to the invariant chain (Ii). This binding has two main functions: (1) it blocks the peptide binding cleft of class-II molecules so that they do not bind peptides in the ER; and, (2) it routes class-II molecules to the endosome, where it encounters proteins from the extracellular fluid or proteins from the cell surface that were taken up by endocytosis. While the endosomes mature into lysosomes, the pH is reduced and proteases are activated, leading to fragmentation of endosomal proteins. The Ii chain is also degraded, except for the class-II associated Ii chain peptide (CLIP), which is located in the cleft [25]. The HLA-DM help empty CLIP from the class-II cleft and enhance the loading of peptides into the empty cleft [54].

Both HLA (MHC) class-I and class-II molecules bind peptides in a cleft formed by two α-helices that rest on the floor of a β-pleated sheet [7, 8, 56]. The class-I cleft is closed at both ends and accommodates peptides that are 8–10 amino acids (aa) long [49], whereas the class-II cleft is open at both ends and, therefore, accommodates longer peptides [49]. Crystallography of both class-I and class-II peptide complexes has shown that the HLA cleft binds peptides by forming hydrogen bonds to the peptide backbone and by sequestrating some aa side chains (anchor aa) inside particular "pockets" deep in the binding cleft [11, 23, 38, 56].

One important feature of HLA (MHC) genes and, thus, of HLA (MHC) molecules is their extensive genetic polymorphism. A large number of allelic variants of HLA genes exist [10]; the number of currently identified allelic variants of genes encoding HLA molecules and molecules involved in antigen processing is shown in Fig. 2. The polymorphic aa are mainly located around the peptide binding cleft, where allelic variants of HLA molecules have different pockets [49]. It follows that a given HLA (MHC) molecule may only bind those peptides that have aa side chains that fit into these pockets. The positions and types of residue that anchor a peptide to a particular cleft determines the peptide-binding motif of that HLA (MHC) molecule. Thus, different HLA (MHC) molecules bind and present different sets of peptides to T cells [49]. The HLA molecules therefore contribute to the immune-response repertoire of the individual.

The extensive genetic polymorphism of HLA (MHC) ensures that there will always be individuals with HLA (MHC) molecules that are able to bind peptides from a given microorganism. This makes it much less likely that microrganisms will evade the immune system by evolving proteins that are unable to bind to any MHC molecule of that species.

5 T Cell Recognition of Foreign Peptides Presented by Self-HLA

T cells are able to recognize foreign (non-self) peptides complexed with self-HLA (MHC) molecules. The peptide/HLA complex is recognized by the T cells by its specific TCR. The majority (>90%) of T cells bear a TCR$\alpha\beta$ receptor, which is a heterodimer composed of an α- and β-chain polypeptide. The remaining T cell population expresses the heterodimeric TCR$\gamma\delta$ receptor, the function of which is less well known than that of the TCR$\alpha\beta$. Recent crystallography of the trimolecular complex, TCR$\alpha\beta$–peptide–HLA class I, has shown that the so-called hypervariable complementary determining region (CDR) loops of the TCR make contact with both the peptide and the HLA [22].

The TCRα and β variable regions are encoded by randomly assembled germ line V, D (only in β) and J gene segments, and by non-germ line *N* segments (reviewed in [19]). During the development from bone marrow stem cells, the maturing T cell population is able to generate a very large number of TCR variants, which are able to recognize an enormous variety of peptide/HLA complexes. During maturation in the thymus, however, only those T cells that are able to recognize peptides presented by self-HLA molecules are positively selected. T cells that generate TCRs that are only able to recognize peptides presented by foreign-HLA molecules are useless to the body and, consequently, are not selected; they die of neglect (reviewed in [64]).

HLA molecules are general peptide receptors that do not distinguish between peptides derived from foreign or self-proteins. They therefore bind all peptides that, in form and charge, fit into their cleft. Positively selected T cells are therefore able to recognize self and non-self (foreign) peptides alike. T cells that are able to recognize self-peptides bound to self-HLA molecules are potentially autoreactive and this may be dangerous. They are therefore deleted during maturation in the thymus (negative selection) [32] or are otherwise inactivated or silenced in the periphery (reviewed in [1]).

Immune responses may be divided into three phases: recognition, activation and effector mechanisms. The same is the case for alloimmune responses.

6 Allorecognition

6.1 Recognition by T Cells

Recipient T cells are able to perceive allogeneic donor cells and tissue in two different ways. First, they may recognize foreign peptides derived from proteins of the transplant and presented by the self-HLA molecules of recipient antigen-presenting cells (APC). This is called indirect allorecognition (Fig. 3). Second, they

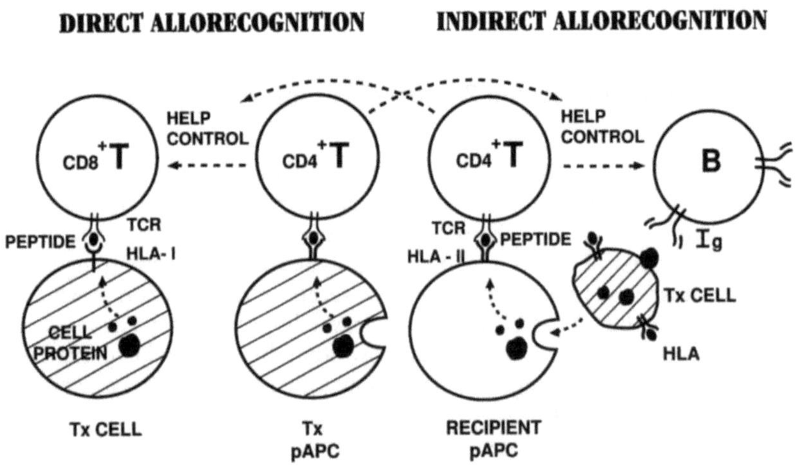

Fig. 3. An overview of indirect and direct allorecognition. CD4+ T cells (*T*) involved in indirect recognition are able to recognize material from transplanted (*Tx*) cells (*cross-hatched*) after the transplanted cells – or fragments from them – are taken up, processed and displayed by a host professional antigen-presenting cell (pAPC). B cells (*B*) may perceive foreign molecules directly on the transplanted cell. CD4+ T cells and CD8+ T cells involved in direct allorecognition recognize the peptide/HLA complexes on transplanted cells

may directly recognize foreign (i.e. non-self) HLA molecules or (usually foreign) peptide/HLA complexes that are present on the transplanted cells (Fig. 4). This is called direct allorecognition (reviewed in [4]).

6.1.1 Indirect Allorecognition

Indirect allorecognition is the result of T cell recognition of foreign allogeneic proteins in the same way as T cells recognize foreign proteins derived from bacteria or viruses. Here, self-HLA molecules present peptides derived from allogeneic proteins in the graft. Alloreactive T cells will only indirectly recognize foreign peptides that have one or more polymorphic aa substitutions; this makes them different from the corresponding self-peptides, since T cells that recognize self-peptides bound by self-HLA molecules are deleted or silenced. Among the prominent peptides eluted from HLA (MHC) molecules are those from other HLA (MHC) molecules, many of which are from polymorphic regions of the HLA (MHC) [15, 49, 63]. It follows that, when donor and recipient are HLA different, recipient T cells will be confronted with many foreign HLA-derived peptides, against which the individual is not tolerant. In contrast, when donor and recipient are HLA identical, recipient T cells cannot react with peptide fragments from donor HLA-molecules, since they would be tolerant to such identical-to-self HLA peptides. In HLA-identical donor-recipient combinations, indirect allorecognition can thus only involve recognition of other polymorphic, non-HLA peptides, such as those from polymorphic regions of Ig molecules. Since the HLA molecules are by far the most polymorphic proteins in the body, indirect allorecognition will involve far more T cells in HLA-different than in HLA-identical donor-recipient combinations.

6.1.2 Direct Allorecognition

T cells can also directly recognize peptide/HLA complexes on donor cells, where the HLA molecules are of donor origin [52]. Since T cells that recognize peptides bound by non-self HLA molecules are not positively selected in the thymus (see 5), it is surprising that there is such a large number of T cells that recognize non-self HLA-molecules. The reason seems to be that the TCR of those T cells that directly recognize a complex of foreign peptides in self-HLA molecules often may also recognize peptides in foreign-HLA molecules, due to structural similarities, i.e. cross-reactions. Thus, it has been shown that T cell clones that are able to recognize a particular complex of a viral peptide bound by a self-HLA molecule, may also recognize a complex of another peptide in a non-self HLA molecule [12, 26]. Since TCRs recognize the three-dimensional structure of peptide/HLA complexes, the topography of the two different complexes recognized by the same TCR must be similar. In addition, the TCR of some T cells may also directly recognize donor allogeneic HLA molecules, independently of the bound peptide [55].

When recipient T cells are confronted with foreign, donor HLA molecules, they are confronted with a large number of foreign peptide/HLA complexes against which self-tolerance has not been generated. The reason is that tolerance is only generated towards self-peptides bound by self-HLA molecules and not towards the different sets of peptides bound by non-self HLA molecules. In HLA-disparate donor-recipient combinations, direct allorecognition will therefore involve many different T cells. In contrast, when donor and recipient are HLA-identical, donor and recipient HLA molecules will contain overlapping sets of peptides, against most of which self tolerance has been generated. In such cases, therefore, direct allorecognition will only involve a few T cells.

6.2 Recognition by Natural Killer (NK) Cells

Natural killer (NK) cells are a particular type of killer cell, which uses receptors other than TCR to recognize ligands. Their activation is inhibited by killer inhibitor receptors (KIR), which recognize self-HLA class-I molecules (reviewed in [41]). Thus, NK cells become killers only when they recognize cells that have strongly downregulated the expression of one or several self-HLA class-I molecules, for example, because of malignant transformation or viral infection. Similarly, NK cells may recognize and kill allogeneic cells that do not express the appropriate self-HLA molecules. Thus, NK cells work in a complementary fashion to T cells; whereas T cells recognize foreign HLA molecules, NK cells recognize the absence of self-HLA molecules.

6.3 Recognition by B Cells

B cells can, by means of their Ig receptors, recognize epitopes on foreign HLA molecules directly [21]. When an individual is exposed to foreign HLA molecules during or after pregnancy, blood transfusion or transplantation, B cells may recognize the non-self polymorphic sites on the foreign HLA molecules directly (Fig. 3). Sometimes, after allotransplantation, B cells may also recognize foreign molecules other than HLA.

7 Alloactivation

7.1 Alloactivation of T Cells

Alloreactive T cells are most probably activated in the same way as other antigen-specific T cells. When a TCR binds with sufficient avidity to the allogeneic peptide/HLA complex, and the CD4 (in helper T cells) or CD8 (in cytotoxic T cells) co-receptors bind to a constant part of HLA (MHC) class-II or HLA (MHC) class-I molecules, respectively, an activation signal – signal 1 – is transduced into the T cell. Signal 1 is necessary, but not sufficient, to activate T cells. Rather, signal 1 alone may cause ignorance or anergy (Fig. 4a). Other signals – collectively called signal 2 – are also necessary for activation (Fig. 4b). The most important signal 2

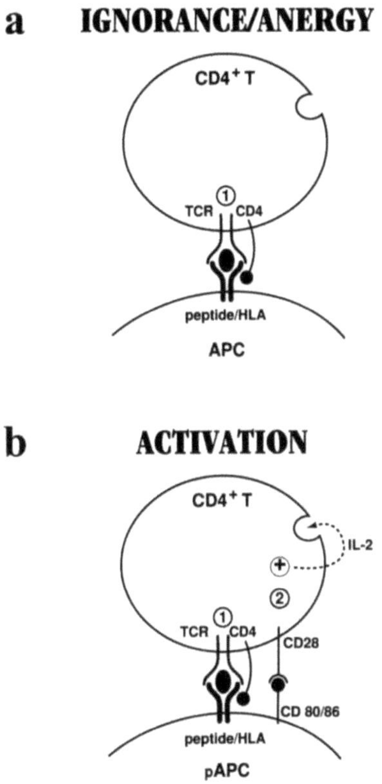

Fig. 4. a CD4+ T cells, which bind only the peptide/human leucocyte antigen (*HLA*) complex and not other accessory molecules, will get only signal-1, will ignore the foreign peptide/HLA or will be made anergic

b CD4+ T cells which, by their T cell receptors (*TCR*), recognize the peptide/HLA complex and with their CD28 molecule bind to the CD80 and/or CD86 molecules, get both signal-1 and signal-2. This will lead to a positive (+) signal for synthesis of interleukin-2 (IL-2), which binds to the IL-2 receptor. Consequently, the T cell will be fully activated and will proliferate. *pAPC*, professional antigen-presenting cells

results from the binding of constitutively expressed CD28 molecules on T cells to CD80 (former B7–1) or CD86 (former B7–2) molecules on APC [38]. Other T cell surface molecules, such as CD40 ligand (gp39) [35], CD2 [30], LFA1 and ICAM1 [28], contribute to signal 2. Following signals 1 and 2, the T cell is fully activated so that the genes encoding lymphokines and lymphokine receptors are transcribed and translated. Cells that bear the ligands (surface molecules) required for the induction of both signals 1 and 2 in the T cell, are often called professional antigen-presenting cells (pAPC).

Following activation of T cells, pAPC also express CTLA-4 which, like CD28, binds to CD80 and CD86 – albeit, at a much higher affinity. Whereas CD28 contributes to signal 2, CTLA-4 inhibits the activation of T cells by competing for the CD28-CD80/CD86 interaction and by transducing a negative signal that can inhibit T cell activation (Fig. 5a) (reviewed in [9]). This prevents over-expansion of activated T cells.

Activation of T cells also leads to activation of another receptor, Fas (CD95), as well as its ligand, Fas ligand (FasL) (reviewed in [1]). The interaction of Fas with FasL (on the same cell or on neighbouring T cells) induces programmed cell death (apoptosis) of the T cell (Fig. 5b). This provides an additional mechanism to control the expansion of activated T cells and, thus, secures self-tolerance.

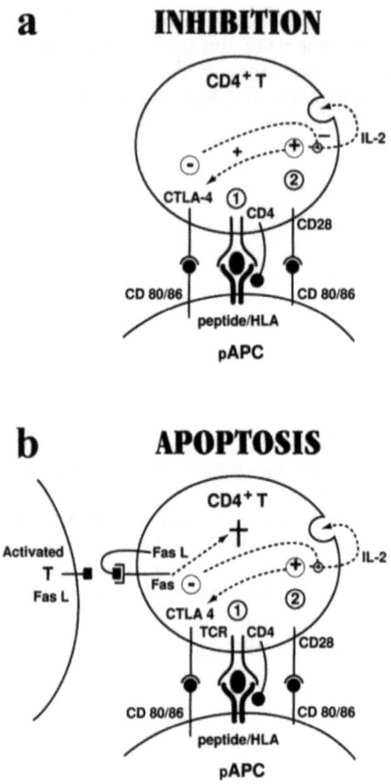

Fig. 5. **a** Activated CD4+ T cells will also express the CTLA-4 molecule. This binds to the CD80/CD86 on the surface of the professional antigen presenting cells (*pAPC*); the interaction leads to a negative (–) signal to T cells, thereby preventing over-expansion of activated T cells

b CD4+ T cells will, upon activation, start to express FasL, which may bind to Fas (CD95) on the same or neighbouring T cells. This will result in an apoptotic death cell programme for the T cells

When CD4+ T cells are fully activated, they secrete interleukin (IL)-2 which binds to IL-2 receptors on CD4+ and CD8+ T cells and, thereby, induces proliferation (clonal expansion) of these cells. CD4+ T cells may differentiate into T helper-1 (Th1) cells or T helper-2 (Th2) cells (reviewed in [57]).

7.2 Alloactivation of B Cells

B cells require help from T cells during activation, proliferation and differentiation into antibody-secreting plasma cells. An HLA-specific B cell is activated if its Ig receptor recognizes a polymorphic site on foreign HLA molecules, i.e. a B cell epitope, and if, at the same time, it receives help from T cells that recognize the same foreign HLA molecule. This may be the result of the internalization of the foreign HLA molecule by B cells, followed by the presentation to T cells (by B cells) of peptides from the HLA molecule. The "help" received from the lymphokines produced by the T cells induces proliferation, isotype switching and differentiation of the B cells [21, 56].

8 Effector Mechanisms in Allograft Rejection

Alloreactive T cells seem to recognize foreign peptide/HLA complexes in the same way as do ordinary antigen-specific T cells [26]. As such, the induced immune reactions are the same as those induced by T cells that recognize an infectious agent; nevertheless, this effect is often stronger due to the fact that more alloreactive T cells are initially activated. As many as 1–10% of T cells in an individual may respond to allogeneic cells [52]. When alloreactive $CD4^+$ T cells are activated on recognition of foreign peptide HLA class-II complexes, they proliferate and, at the same time, secrete a set of lymphokines. The alloreactive T cell response seems to be mainly a Th1 response, with secretion of IL-2 and interferon gamma (IFNγ) [17]. IL-2 helps sustain proliferation of $CD4^+$ and $CD8^+$ T cells that, at the same time, have recognized foreign peptide/HLA class-I complexes. If NK cells fail to recognize their self-HLA molecules on the grafted cells, they are no longer inhibited and may kill the target cells. IFNg both enhances the killer potential of $CD8^+$ T and NK cells and, at the same time, activates monocytes/macrophages. Help from alloreactive $CD4^+$ Th2 cells is required for activation, proliferation and differentiation of alloreactive HLA-specific B cells into plasma cells that – depending on the HLA disparity between donor and recipient – secrete antibodies specific for donor HLA class-I or class-II molecules.

In concert, alloreactive T cells, NK cells and monocytes attack and destroy the foreign cells, both by direct cell-to-cell contact killing and by means of the cytokines secreted by these effector cells. Direct cell-to-cell contact T cell killing is mediated through two different pathways. One is the expression of FasL by activated T cells, which binds to Fas (CD95) on the target cell and, thereby, initiates apoptosis of the target cells [29]. Alternatively, $CD8^+$ T or NK cells may kill target cells through the release of granules containing perforin and granzyme B [27, 53]. Perforin will, at high concentrations, lead to necrotic cell death by forming lytic pores in the cell membrane, or it may, at sublytic concentrations, enhance the transport of granzyme B into the cell where it initiates the same apoptotic program as is mediated by Fas. During acute rejection, $CD4^+$ T cells, $CD8^+$ T cells, NK cells, B cells and monocytes/macrophages are all present among graft-infiltrating cells, and cytotoxic effector and delayed-type hypersensitivity are the main mechanisms responsible for the destruction of the graft [18, 34, 58].

If HLA antibodies are produced after transplantation, they may – in concert with T and NK cells – contribute to rejection by activating complement or by enhancing the activity of monocytes and NK cells, via enhanced Fc receptor binding to target cells. By binding directly to HLA molecules of the endothelial cells of the transplant and activating complement, preformed HLA class-I specific antibodies, present at the time of transplantation, may cause hyperacute rejection, thereby leading to vascular occlusion. Preformed HLA class-II specific antibodies do not usually cause hyperacute rejection, probably because endothelial cells express low levels of HLA class-II molecules.

9 Role of HLA Matching in Clinical Transplantation

Because of the major role that donor HLA molecules play in indirect and direct allorecognition (see Sects. 6.1.1, 6.1.2), it is to be expected that the immune response against allografts will be less prominent if the number of foreign donor HLA molecules is kept at a minimum. Many large multi-centre studies show a significant beneficial effect of HLA matching, both in kidney and heart transplantation [46–48]. In our own work, a prominent effect of matching for HLA molecules can be demonstrated. The projected half-life of zero HLA haplotype mismatched kidneys are twice as long as that seen for one- and two-HLA haplotype mismatched living donor kidney transplantation (Torbjørn Leivestad, personal communication). In our own work, matching for HLA-DR molecules appears to be more important than matching for HLA-A and -B molecules. Thus, the projected half-life of zero HLA-DR mismatched cadaveric kidneys is twice as long (12 years vs 6 years) as that observed for one and two HLA-DR mismatched cadaveric kidneys (Fig. 6), whereas few or no significant effects are found for the matching of HLA-A and -B molecules [50]. Because of the role that $CD4^+$ T cells (Th1) play in controlling and amplifying allograft responses [57], and because $CD4^+$ T cells recognize peptide/ HLA class-II complexes, the predominant role of DR molecules for graft survival is not unexpected.

Whereas matching for HLA molecules benefits organ transplantation, careful HLA matching is necessary for bone-marrow transplantation [3]. The risk of graft failure due to HVG responses and serious GVH disease is greatly increased if HLA-mismatched bone-marrow transplantation is performed. Since only approximately 25% of patients have an HLA-identical sibling donor, a large number of

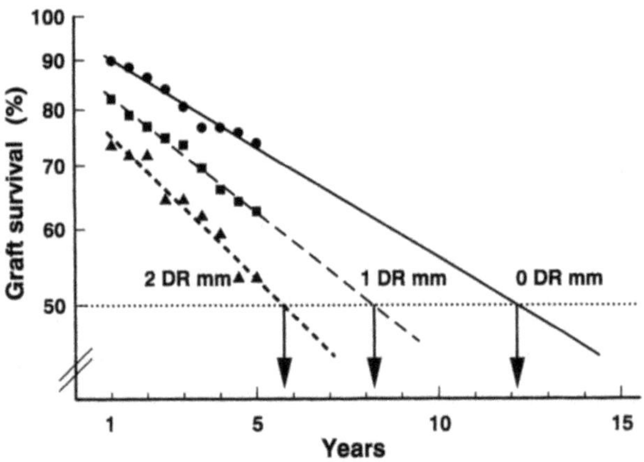

Fig. 6. Influence of human leucocyte antigen (HLA)-DR matching on cadaveric renal graft survival (data from The Transplantation Unit, The National Hospital, Oslo). Observed survival rate of first cadaveric grafts mismatched (*mm*) for zero (*filled circles; n*= 231), one (*filled squares; n*=370) or two (*filled triangles; n*=64) DR antigens (DR1–10) and was given to non-sensitised recipients. The respective regression lines and projected half-lives of the groups are indicated

bone-marrow registries have been established around the world. With the help of these bone-marrow registries, which have now HLA-typed approximately 4 million individuals (mainly Caucasoids), it is possible to obtain acceptably HLA-matched bone marrow from unrelated donors for the majority of Caucasoid patients that are in need of transplantation.

High quality HLA typing is required for both organ and bone-marrow transplantation. The advent of simpler and more precise methods for typing HLA molecules (serological typing) [62] and HLA genes (genomic typing) [44] has probably also contributed to the better graft survival and the reduced incidence of complications that are seen for well-matched HLA combinations.

10 Specific Immunosuppression

Immunosuppressive treatment of allograft recipients is always necessary, even following transplantation of organs or bone marrow from an HLA-identical sibling. Despite the fact that basic research has brought us detailed insight on how immunocompetent cells perceive foreign molecules and cells specifically, the immunosuppressive therapy given after transplantation is not efficient enough to prevent immune-mediated graft loss in a sizeable number of patients. Moreover, the immunosuppressive therapy given is non-specific and will therefore affect useful and unwanted immune reactions alike. Transplant patients are therefore prone to serious infection and cancer [40]. In addition, the long-term treatment of patients with immunosuppressive drugs results in non-immunologically mediated side effects, often affecting the CNS, the cardiovascular system and the skin. If alloimmune responses towards the donor could be specifically suppressed, without affecting immune responses to other antigens and without side effects, patient mortality and morbidity could be reduced significantly. This is the aim behind attempts at inducing specific non-responsiveness or tolerance towards donor cells.

10.1 Mechanisms for the Maintenance of Specific Non-responsiveness: Tolerance

Normally, we are tolerant to self. Several mechanisms are responsible for this (only T cell mechanisms will be briefly discussed here).

10.1.1 Deletion

T cells that recognize complexes of self-peptide/HLA molecules are deleted by negative selection during development in the thymus (reviewed in [39]). T cells may also be deleted in the periphery during an immune response. Thus, following activation, T cells will express Fas, which, upon binding to FasL on the same T cell or on another activated T cell, will initiate apotosis (see Sect. 7.1).

10.1.2 Ignorance and Anergy

T cells will ignore peptide/HLA complexes recognized by their TCR unless, at the same time, they receive a second signal through their CD28 co-receptors (see Sect. 7.1). If the ligand for the CD28 is not present on the APC presenting the peptide/HLA complex, the T cells ignore it or else are made specifically non-responsive or anergic [51] (Fig. 4a). Most parenchymal cells do not normally express the CD80 or CD86 molecules. This may be an important mechanism that underlies non-responsiveness to tissue-specific peptide/HLA complexes, which are not present in the thymus, and thereby do not induce negative selection of T cells having the corresponding TCR.

10.1.3 Suppression

Specific non-responsiveness may be transferred from one animal to another. For example, the transfer of lymphocytes from an animal successfully transplanted with an allograft to a syngeneic host will modify the alloimmune responses if the latter animal is then transplanted [24, 33]. The nature of such suppressor lymphocytes is not known.

10.2 Methods for the Induction of Specific Non-responsiveness: Tolerance

Billingham, Brent and Medawar showed, in their seminal 1953 study, that long-lasting tolerance to skin allografts could be achieved in mice inoculated in utero with a mixture of donor strain cells, including spleen cells [6]. It has been suggested that long-lasting tolerance to an allograft can only be maintained if donor antigens persist in the recipient. By establishing a mixed donor/recipient multilineage chimerism, Kawai and co-workers have succeeded in inducing tolerance to renal allografts in primates [31].

Induction of reduced- or non-responsiveness to grafts has been attempted by performing donor-specific blood transfusions prior to transplantation. This has been shown to be efficient in both experimental [16, 20] and clinical studies [46]. However, after cyclosporin was introduced, the effect of donor-specific transfusions became much less pronounced [2]. It is possible that this failure to induce tolerance in patients treated with cyclosporin is due to its blocking effect on the TCR-mediated signal 1 that is required for the induction of anergy. In a large number of patients, donor-specific transfusions also induce the production of antibodies to donor HLA antigens, so that the patient cannot be transplanted from the selected donor. For these reasons, donor-specific transfusions have largely been abandoned in clinical transplantations.

Blocking of CD4 molecules has proved efficient for inducing tolerance in animal models [5]. Treatment of patients with the same antibody, however, has not been sufficient to improve graft survival. In animal models, specific non-responsiveness has also been induced in alloreactive T cells by blocking the C28-CD80/CD86 interaction via treatment with soluble CTLA-4 molecules (CTLA-4 Ig) which bind with high affinity to CD80/CD86 [37, 61].

Since the allograft reaction is mostly Th1 mediated [17, 42, 60] and a reciprocal relationship exists between Th1 and Th2 immune responses, the induction of graft tolerance has been attempted by the administration of Th2 derived lymphokines, IL-4 and IL-10. Such treatment strategies have, however, not been successful. The reason may be that lymphokines other than IL-2 and IFNγ may mediate graft rejection. Thus, it has been shown that both IL-2- and IFNγ knock out (KO) mice are fully capable of rejecting allografts, albeit the graft survival rates are moderately increased compared with that observed in wild-type mice [57]. That there is a redundancy of cytokines able to mediate graft rejection makes it less likely that the induction of regulator cells will be an efficient means to prevent graft loss.

11 Concluding Remarks

Rapid development in basic biomedical research has provided good insight into the mechanisms of allorecognition. The role of the HLA molecules of donor cells, both in indirect and direct allorecognition, has been amply demonstrated, and explains the impact of HLA matching in clinical organ and bone marrow transplantation. Unfortunately, this insight has not yet led to methods that induce donor-specific non-responsiveness in patients. We still have to rely on non-specific immunosuppressants, which may have serious side-effects both on the immune system and on other organs. It is hoped and expected that the insight basic research has brought us may soon pay off in the form of more specific ways to secure long-lasting graft survival with fewer side effects.

References

1. Abbas AK (1996) Die and let live: eliminating dangerous lymphocytes. Cell 84: 655–657
2. Ahmed Z, Terasaki PI (1991) Effect of transfusions. Clin Transpl 305–312
3. Anasetti C, Beatty PG, Storb R, Martin PJ, Mori M, Sanders JE, Thomas ED, Hansen JA (1990) Effect of HLA incompatibility on graft-versus-host disease, relapse, and survival after marrow transplantation for patients with leukemia or lymphoma. Hum Immunol 29: 79–91
4. Auchincloss H, Jr., Sultan H (1996) Antigen processing and presentation in transplantation. Curr Opin Immunol 8: 681–687
5. Benjamin RJ, Waldmann H (1986) Induction of tolerance by monoclonal antibody therapy. Nature 320: 449–451
6. Billingham RE, Brent L, Medawar PB (1953) Actively acquired tolerance of foreign cells. Nature 172: 603–606
7. Bjorkman PJ, Saper MA, Samraoui B, Bennett WS, Strominger JL, Wiley DC (1987a) Structure of the human class I histocompatibility antigen, HLA-A2. Nature 329: 506–512
8. Bjorkman PJ, Saper MA, Samraoui B, Bennett WS, Strominger JL, Wiley DC (1987b) The foreign antigen binding site and T cell recognition regions of class I histocompatibility antigens. Nature 329: 512–518
9. Bluestone JA (1997) Is CTLA-4 a master switch for peripheral T cell tolerance. J Immunol 158: 1989–1993
10. Bodmer JG, Marsh SGE, Albert ED, Bodmer WF, Bontrop RE, Charron D, Dupont B, Erlich HA, Fauchet R, Mach B, Mayr WR, Parham P, Sasazuki T, Schreuder GMT, Strominger, Svejgaard A, Terasaki PI (1996) Nomenclature for factors of the HLA system. Hum Immunol 53: 9–128

11. Brown JH, Jardetzky TS, Gorga JC, Stern LJ, Urban RG, Strominger JL, Wiley DC (1993) Three-dimensional structure of the human class II histocompatibility antigen HLA-DR1. Nature 364: 33–39
12. Burrows SR, Khanna R, Burrows JM, Moss DJ (1994) An alloresponse in humans is dominated by cytotoxic T lymphocytes (CTL) cross-reactive with a single Epstein-Barr virus CTL epitope: implications for graft-versus-host disease. J Exp Med 179: 1155–1161
13. Busch R, Mellins ED (1996) Developing and shedding inhibitions: how MHC class II molecules reach maturity. Curr Opin Immunol 8: 51–58
14. Campbell RD, Trowsdale J (1993) Map of the human MHC. Immunol Today 14: 349–352
15. Chicz RM, Urban RG, Gorga JC, Vignali DA, Lane WS, Strominger JL (1993) Specificity and promiscuity among naturally processed peptides bound to HLA-DR alleles. J Exp Med 178: 27–47
16. Dallman MJ, Wood KJ, Morris PJ (1987) Specific cytotoxic T cells are found in the non-rejected kidneys of blood-transfused rats. J Exp Med 165: 566–571
17. Dallman MJ, Larsen CP, Morris PJ (1991) Cytokine gene transcription in vascularised organ grafts: analysis using semiquantitative polymerase chain reaction. J Exp Med 174: 493–496
18. Dallman MJ (1995) Cytokines and transplantation: Th1/Th2 regulation of the immune response to solid organ transplants in the adult. Curr Opin Immunol 7: 632–638
19. Davis MM, Bjorkman PJ (1988) T-cell antigen receptor genes and T-cell recognition. Nature 334: 95–402
20. Fabre JW, Morris PJ (1972) The effect of donor strain blood pretreatment on renal allograft rejection in rats. Transplantation 14: 608–617
21. Fabre JW (1996) The role of polymorphic donor peptides in allograft recognition and rejection. Immunol Rev 154: 21–43
22. Garboczi DN, Ghosh P, Utz U, Fan QR, Biddison WE, Wiley DC (1996) Structure of the complex between human T-cell receptor, viral peptide and HLA-A2. Nature 384: 134–141
23. Garrett TP, Saper MA, Bjorkman PJ, Strominger JL, Wiley DC (1989) Specificity pockets for the side chains of peptide antigens in HLA-Aw68. Nature 342: 692–696
24. Gassel HJ, Hutchinson IV, Engemann R, Morris PJ (1992) The role of T suppressor cells in the maintenance of spontaneously accepted orthotopic rat liver allografts. Transplantation 54: 1048–1053
25. Ghosh P, Amaya M, Mellins E, Wiley DC (1995) The structure of an intermediate in class II MHC maturation: CLIP bound to HLA-DR3. Nature 378: 457–462
26. Gjertsen HA, Lundin KE, Hansen T, Thorsby E (1993) T cells specific for viral antigens presented by HLA-Dw4 recognize DR13 on allogeneic cells: a possible mechanism for induction of rejection. Transplant Proc 25: 70–71
27. Heusel JW, Wesselschmidt RL, Shresta S, Russell JH, Ley TJ (1994) Cytotoxic lymphocytes require granzyme B for the rapid induction of DNA fragmentation and apoptosis in allogeneic target cells. Cell 76: 77–987
28. Isobe M, Yagita H, Okumura K, Ihara A (1992) Specific acceptance of cardiac allograft after treatment with antibodies to ICAM-1 and LFA-1. Science 255: 1125–1127
29. Kagi D, Vignaux F, Ledermann B, Burki K, Depraetere V, Nagata S, Hengartner H, Golstein P (1994) Fas and perforin pathways as major mechanisms of T cell-mediated cytotoxicity. Science 265: 528–530
30. Kapur S, Khanna A, Sharma VK, Li B, Suthanthiran M (1996) CD2 antigen targeting reduces intragraft expression of mRNA-encoding granzyme B and IL-10 and induces tolerance. Transplantation 62: 249–255
31. Kawai T, Cosimi AB, Colvin RB, Powelson J, Eason J, Kozlowski T, Sykes M, Monroy R, Tanaka M, Sachs DH (1995) Mixed allogeneic chimerism and renal allograft tolerance in cynomolgus monkeys. Transplantation 59: 256–262
32. Kisielow P, Teh HS, Bluthmann H, von Boehmer H (1988) Positive selection of antigen-specific T cells in thymus by restricting MHC molecules. Nature 335: 730–733
33. Knoop M, Pratt JR, Hutchinson IV (1994) Evidence of alloreactive T suppressor cells in the maintenance phase of spontaneous tolerance after orthotopic liver transplantation in the rat. Transplantation 57: 1512–1515
34. Krensky AM, Weiss A, Crabtree G, Davis MM, Parham P (1990) T-lymphocyte-antigen interactions in transplant rejection. N Engl J Med 322: 510–517

35. Larsen CP, Alexander DZ, Hollenbaugh D, Elwood ET, Ritchie SC, Aruffo A, Hendrix R, Pearson TC (1996) CD40-gp39 interactions play a critical role during allograft rejection. Suppression of allograft rejection by blockade of the CD40-gp39 pathway. Transplantation 61: 4–9
36. Lehner PJ, Cresswell P (1996) Processing and delivery of peptides presented by MHC class I molecules. Curr Opin Immunol 8: 59–67
37. Lenschow DJ, Zeng Y, Thistlethwaite JR, Montag A, Brady W, Gibson MG, Linsley PS, Bluestone JA (1992) Long-term survival of xenogeneic pancreatic islet grafts induced by CTLA4Ig. Science 257: 789–792
38. Lenschow DJ, Walunas TL, Bluestone JA (1996) CD28/B7 system of T cell costimulation. Annu Rev Immunol 14: 233–258
39. Madden DR, Gorga JC, Strominger JL, Wiley DC (1992) The three-dimensional structure of HLA-B27 at 2.1 A resolution suggests a general mechanism for tight peptide binding to MHC. Cell 70: 1035–1048
40. Marrack P, Kappler J (1988) The T-cell repertoire for antigen and MHC. Immunol Today 9: 308–315
41. McGregor JM, Proby CM, Leigh IM (1996) Virus infection and cancer risk in transplant recipients. Trends Microbiol 4: 2–3
42. Moretta A, Bottino C, Vitale M, Pende D, Biassoni R, Mingari MC, Moretta L (1996) Receptors for HLA class-I molecules in human natural killer cells. Annu Rev Immunol 14: 619–648
43. O'Connell PJ, Pacheco-Silva A, Nickerson PW, Muggia RA, Bastos M, Kelley VR, Strom TB (1993) Unmodified pancreatic islet allograft rejection results in the preferential expression of certain T cell activation transcripts. J Immunol 150: 1093–1104
44. Olerup O, Zetterquist H (1992) HLA-DR typing by PCR amplification with sequence-specific primers (PCR-SSP) in 2 hours: an alternative to serological DR typing in clinical practice including donor-recipient matching in cadaveric transplantation. Tissue Antigens 39: 225–235
45. Opelz G, Sengar DP, Mickey MR, Terasaki PI (1973) Effect of blood transfusions on subsequent kidney transplants. Transplant Proc 5: 253–259
46. Opelz G, Mytilineos J, Scherer S, Dunckley H, Trejaut J, Chapman J, Fischer G, Fae I, Middleton D, Savage D, et al (1993) Analysis of HLA-DR matching in DNA-typed cadaver kidney transplants. Transplantation 55: 782–785
47. Opelz G, Wujciak T (1994) The influence of HLA compatibility on graft survival after heart transplantation. The collaborative transplant study N Engl J Med 330: 816–819
48. Opelz G, Scherer S, Mytilineos J (1997) Analysis of HLA-DR split-specificity matching in cadaver kidney transplantation: a report of the collaborative transplant study. Transplantation 63: 7–59
49. Rammensee HG, Friede T, Stevanoviic S (1995) MHC ligands and peptide motifs: first listing. Immunogenetics 41: 178–228
50. Reisæter AV, Leivestad T, Fauchald P, Brekke I, Thorsby E (1997) The influence of matching for HLA-DR in first cadaveric renal transplantation in non-sensitized recipients. Transplant Proc 29: 3099–3100
51. Schwartz RH (1996) Models of T cell anergy: is there a common molecular mechanism? J Exp Med 184: 1–8
52. Sherman LA, Chattopadhyay S (1993) The molecular basis of allorecognition. Annu Rev Immunol 11: 385–402
53. Shresta S, MacIvor DM, Heusel JW, Russell JH, Ley TJ (1995) Natural killer and lymphokine-activated killer cells require granzyme B for the rapid induction of apoptosis in susceptible target cells. Proc Natl Acad Sci U S A 92: 5679–5683
54. Sloan VS, Cameron P, Porter G, Gammon M, Amaya M, Mellins E, Zaller DM (1995) Mediation by HLA-DM of dissociation of peptides from HLA-DR. Nature 375: 802–806
55. Smith PA, Brunmark A, Jackson MR, Potter TA (1997) Peptide-independent recognition by alloreactive cytotoxic T lymphocytes (CTL). J Exp Med 185: 1023–1033
56. Stern LJ, Brown JH, Jardetzky TS, Gorga JC, Urban RG, Strominger JL, Wiley DC (1994) Crystal structure of the human class II MHC protein HLA-DR1 complexed with an influenza virus peptide. Nature 368: 215–221

57. Strom TB, Roy-Chaudhury P, Manfro R, Zheng XX, Nickerson PW, Wood K, Bushell A (1996) The Th1/Th2 paradigm and the allograft response. Curr Opin Immunol 8: 688–693
58. Suthanthiran M, Strom TB (1994) Renal transplantation. N Engl J Med 331: 365–376
59. Swain SL (1983) T cell subsets and the recognition of MHC class. Immunol Rev 74: 129–142
60. Takeuchi T, Lowry RP, Konieczny B (1992) Heart allografts in murine systems. The differential activation of Th2-like effector cells in peripheral tolerance. Transplantation 53: 1281–1294
61. Turka LA, Linsley PS, Lin H, Brady W, Leiden JM, Wei RQ, Gibson ML, Zheng XG, Myrdal S, Gordon D, et al (1992) T-cell activation by the CD28 ligand B7 is required for cardiac allograft rejection in vivo. Proc Natl Acad Sci U S A 89: 1102–11105
62. Vartdal F, Gaudernack G, Funderud S, Bratlie A, Lea T, Ugelstad J, Thorsby E (1986) HLA class I and II typing using cells positively selected from blood by immunomagnetic isolation- a fast and reliable technique. Tissue Antigens 28: 301–312
63. Vartdal F, Johansen BH, Friede T, Thorpe CJ, Stevanovic S, Eriksen JE, Sletten K, Thorsby E, Rammensee HG, Sollid LM (1996) The peptide binding motif of the disease associated HLA-DQ (α1* 0501, β1* 0201) molecule. Eur J Immunol 26: 2764–2772
64. von Boehmer H (1994) Positive selection of lymphocytes. Cell 76: 219–228

Infections in Solid Organ Transplant Recipients

C. C. Kibbler

1 Introduction

The spectrum of infection in solid organ-transplant recipients has widened over the last decade as the numbers of patients transplanted and the diversity of procedures undertaken has increased. Some of these infections are only described in a handful of case reports, but for others a pattern has emerged of relatively common infections, occurring at specific times during the post-transplant period (see Table 1 and Fig. 1).

Transplant patients differ from other groups of immunosuppressed patients, such as those with HIV infection, by virtue of the fact that they combine the infective complications of surgery with a variable period of immunosuppression. The regimens needed to maintain the delicate balance between graft rejection and

Current Topics in Pathology
Volume 92, C. L. Berry (Ed.)

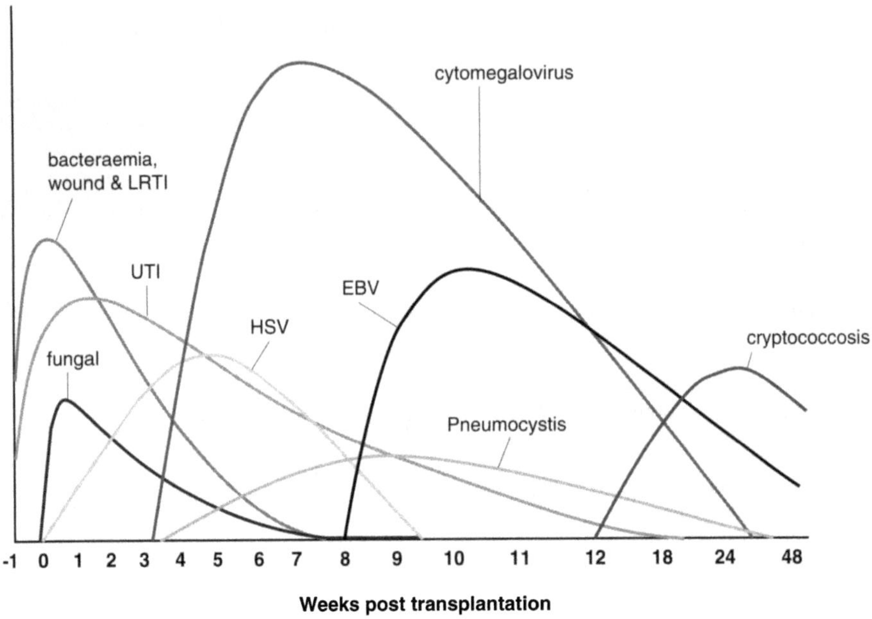

Fig. 1. Sequence of infections following solid organ transplantation

Table 1. Important infectious agents in organ transplant recipients

Infectious agent	
Bacteria • Staphylococci • Streptococci • *Enterobacteriaceae* • Pseudomonads • *Mycobacterium* spp. • *Legionella* spp. • *Clostridium difficile*	Viruses • Herpes simplex • Varicella zoster • Cytomegalovirus • Epstein-Barr virus • Hepatitis B and C • Polyomavirus • Adenovirus
Fungi • *Candida* spp. • *Aspergillus* spp. • Zygomycetes • *Cryptococcus neoformans* • *Pneumocystis carinii*	Protozoa/parasites • *Toxoplasma gondii* • *Strongyloides stercoralis*

infection risk have been refined over the past two decades and newer agents are being introduced. It is likely that the use of these will reduce the incidence of opportunistic infections in these patients. Most of the studies on which this review is based were carried out in patients receiving a triple regimen of azathioprine, corticosteroids and cyclosporin.

2 Time Course of Infections

There is a sequence of infectious complications following any transplant procedure. Knowledge of this is helpful in guiding selection of appropriate diagnostic investigations, as well as establishing the appropriate duration of antimicrobial prophylaxis and choice of empirical therapy.

2.1 Infections in the First Month Post-transplant

The most frequent infections in the first month are those associated with the surgical procedure, including those complicating the anastomoses peculiar to the specific transplant procedure, such as the biliary tract in liver transplantation. In addition, some infections are transmitted with the allograft or were present in the recipient prior to transplantation, particularly if undergoing a repeat transplant procedure.

The only common viral infection seen during the first month is that due to herpes simplex virus (HSV) in seropositive patients. This is usually confined to the mucosae, but may occasionally disseminate and involve the graft. Problem infections with this virus have been largely eliminated with the widespread use of acyclovir in seropositive patients.

2.2 Infections Between 1 Month and 6 Months Post-transplant

Organisms typically associated with the immunocompromised host cause infection 1–6 months following transplant, and the risk of infection correlates with the severity of immunosuppression, such that patients requiring therapy for acute rejection episodes are at increased risk. The most numerically important infections at this time are those due to the herpes group of viruses, especially cytomegalovirus (CMV), *Pneumocystis carinii* and other fungi, *Toxoplasma gondii, Nocardia* species and *Listeria monocytogenes.* In addition, patients with other latent infections, such as tuberculosis or histoplasmosis, may suffer reactivation disease during this period.

Patients undergoing repeat transplantation will often have the risk of these infections superimposed upon the earlier infections discussed above.

2.3 Infections Occurring After 6 Months

The recipient is at risk of community acquired infections following discharge from hospital, but these infections usually form the majority of those occurring in the late transplant period. Some patients remain chronically severely immunosupressed because of chronic rejection or chronic infection with CMV; these patients are at continued risk of infection with opportunistic organisms. Otherwise, the only relatively common opportunistic infection occurring at this stage is reactivation of varicella-zoster virus (VZV) infection.

3 Renal Transplantation

3.1 Bacterial Infections

Bacterial infections occur in approximately 50% renal transplant recipients during hospitalisation [1]. The main surgical complications of implanting a transplanted kidney are peri-graft haematoma, urinary leaks from the ureteric anastomosis and lymphocoeles. All of these may be associated with subsequent infection, but the most common is urinary tract infection (UTI). Up to 90% of recipients may suffer this following discharge [2]. Risk factors for UTI include prolonged haemodialysis pre-transplant and prolonged urinary catheterisation. Besides the Enterobacteriaceae, common organisms causing UTI in these patients include enterococci (including vancomycin-resistant enterococci on some units), staphylococci and *Pseudomonas aeruginosa.* Renal stone formation in some patients has been associated with *Corynebacterium urealyticum* which produces urease, changing the pH of the urine. Prophylaxis with co-trimoxazole results in a reduction in the incidence of UTI by at least 50% [3, 4]. Urinary tract infections in these patients are often asymptomatic and may occur with few white cells in the urine. Unsuspected infections may be diagnosed on renal biopsy, performed for presumed rejection.

Bacteria isolated from the graft perfusion fluid vary in their propensity to cause post-transplant infection. Positive cultures have been found in 4–40% of renal transplants, but most have been with gram-positive skin commensal bacteria [5–7]. However, the isolation of the Enterobacteriaceae and *Pseudomonas aeruginosa* correlates with vascular infection and post-operative sepsis [5, 8, 9]

Outbreaks of infections with *Nocardia* species were first described on renal units and renal-transplant recipients remain among the most common single group of patients to suffer these infections, which occur in up to 4%. Patients typically present with respiratory involvement and pulmonary infiltrates, nodules or cavitary lesions on chest radiographs. Dissemination to the central nervous system (in the form of brain abscess, or meningitis) or skin may occur and occasionally joint infections, myocarditis, genital infection and infection of the renal graft itself may be seen [10–12].

The risk of non-enteric *Salmonella* infection is more than 20-fold greater in the renal-transplant recipient than in the normal population and appears to be related to the degree of immunosuppression [13]. Patients may present with bacteraemia or disseminated infection, involving joints, lung abscess, perinephric abscess, lymphocoele infection, soft-tissue infection and infections involving the vascular anastomosis. Such infections may require surgery as well as antimicrobial therapy. Recurrent *Salmonella* infection is common and suggests a latent focus of infection.

Transplant recipients are at increased risk of legionnaire's disease, which typically presents with pulmonary signs and symptoms. The graft itself is not usually involved, although renal impairment is a common multi-system feature of this disease. The risk of this infection should always be considered in transplant recipients as there is a high risk of exposure to the organism during the in-patient phase. A recent survey of transplant units in the UK demonstrated that *Legionella* species could be isolated from the water in approximately 50% [14].

The risk of tuberculosis is up to 50 times greater than in the rest of the local population and the incidence reflects that in the local area. Thus, the incidence in a study from Oxford, UK, was 1.7%, whereas that from a Saudi Arabian study was 3.5% [15]. Other mycobacterial infections occur extremely rarely in the renal-transplant recipient and include infection of skin, lung and bone, and disseminated infection with *M. kansasii, M. chelonae* and *M. xenopi* as well as superficial infection with less virulent species.

3.2 Viral Infections

As in all organ-transplant recipients, CMV infection is common in renal-transplant recipients, occurring in more than 30% in some series. However, CMV disease is less common than in other transplant groups and is symptomatic in less than 10% [16]. Symptomatic infection presents with fever, occasional myalgia and arthralgia, and malaise which may be associated with myelosuppression; the atypical lymphocytosis seen with CMV infection in immunocompetent individuals is rarely detected.

The graft may be affected by a typical glomerular involvement. There is a mononuclear infiltration and an accumulation of fibrillar material in the glomerular capillaries. The endothelial cells themselves may become enlarged or even necrotic [17].

More severe CMV disease tends to occur in primary infection, that is, in the seronegative recipient of an organ from a seropositive donor, and in those with more profound immunosuppression, following therapy for episodes of acute rejection, particularly with antilymphocyte preparations. Other organs may be involved, especially the lung, gastrointestinal tract, biliary system, retina and central nervous system. CMV retinitis presents late and is usually seen in patients with chronic immunosuppression; it is rare when compared with the incidence in HIV-infected patients.

Diagnosis may be made on histological examination, by identification of typical inclusion bodies. Immunohistochemistry or DNA hybridisation techniques may also be used to detect the virus in tissue. However, the diagnosis is usually made by culture, detection of early antigen, antigenaemia or polymerase chain reaction (PCR) [18–21] in blood, urine or other fluids, such as bronchoalveolar lavage fluid.

Therapy of significant CMV disease involves 2–3 weeks of ganciclovir, although retinitis requires longer treatment. Patients with pneumonitis should be given additional intravenous immunoglobulin. Those who are intolerant of ganciclovir or fail to respond can be treated with foscarnet. The mortality of CMV disease is difficult to assess, as patients often die of other associated disease. However, the mortality associated with pneumonitis is high in most transplant groups, with at least 80% of patients dying in those series published prior to the advent of ganciclovir and immunoglobulin therapy.

Epstein-Barr virus (EBV) disease post-transplant is probably underdiagnosed. With most recipients seropositive for the virus, most cases are due to reactivation, but primary infection does occur, usually after the patient has been discharged to the community, and is responsible for more severe disease in most cases. Whilst a mononucleosis syndrome may occur, the most important complication is post-transplant lymphoproliferative disorder (PTLD). This condition represents a spectrum of disease from non-specific hyperplasia through to immunoblastic sarcoma, and the cell type may be monoclonal or polyclonal. Risk of this disease appears to be increased by the use of anti-T cell antibodies, such as OKT3 for the treatment of acute rejection and primary EBV viraemia [22]; the overall incidence is approximately 1%.

The presentation of this condition may follow a mononucleosis syndrome or fever of unknown origin, but may also take the form of gastrointestinal disease, such as bleeding, obstruction and perforation, or central nervous system involvement, and the graft itself is often affected. The virus has also been associated with squamous cell carcinoma and smooth muscle tumours in these patients [23, 24]. Therapy of this condition includes reducing or stopping immunosuppressive therapy and returning the patient to haemodialysis, together with local measures, such as surgery, radiotherapy or systemic chemotherapy.

Before the use of acyclovir prophylaxis, HSV infections (most of which were reactivations of latent infection) caused clinical disease in approximately 50% of seropositive patients. In most cases, this takes the form of oral or labial lesions, although more extensive ano-genital involvement, oesophageal or even disseminated infection may occur. There is no characteristic involvement of the kidney in the renal-transplant recipient. Diagnosis is made by electron microscopy of vesicular fluid, immunofluorescence or culture of fluid or tissue. Treatment is with acyclovir for up to 2 weeks.

VZV infections occur in up to 12% of renal-transplant recipients. Reactivation is typically seen after the first 6 months post-transplant, but primary infection may occur earlier and can cause a life-threatening condition, including haemorrhagic pneumonitis, rash, encephalitis, pancreatitis, disseminated intravascular coagulopathy and hepatitis. Diagnosis is made in the same way as for HSV infection, and treatment is with high-dose acyclovir.

Renal transplantation in patients who are hepatitis B-virus surface-antigen (HBsAg) positive is associated with progression of liver disease, in most cases. The majority of those with severe chronic active hepatitis and up to 60% of those with mild chronic active hepatitis will progress to cirrhosis post-transplant [25]. However, whether this is associated with increased mortality is controversial. Acute hepatitis B following renal transplantation carries the risk of early death in the post-transplant period from liver failure [25]. Hepatitis C virus is a common cause of chronic liver disease in renal-transplant recipients, although little effect on patient or graft survival has been demonstrated [26–28].

Polyomavirus infections have been described in renal transplant recipients. JC virus associated progressive multifocal leucoencephalopathy has occurred rarely and BK virus has been detected in the urine of patients developing urinary-tract abnormalities, such as urethral strictures [29]. The therapy of these infections is still largely ineffective.

Papilloma virus infection carries an increased risk of cervical carcinoma in female renal-transplant recipients [30] and urinary-tract infection is associated with urinary papillomatosis [31].

Adenovirus infection has been associated with haemorrhagic cystitis in these patients, as well as interstitial pneumonitis and disseminated infection. The virus has been transmitted with the graft in some cases [32].

3.3 Fungal Infections

Of all solid organ-transplant recipients, renal-transplant patients are at least risk of fungal infection, with infections occurring in up to 14% in older series, but probably now with an incidence of less than 5% [33]. The majority of infections occur with *Candida* species (mostly *C. albicans*), and these occur in the first 2 months post-transplant. Most involve the renal tract and are catheter related. *Cryptococcus neoformans* and *Aspergillus* species are responsible for approximately 15%, and the rest are caused by *Pneumocystis carinii*, the endemic fungi and the zygomycetes.

The presentation of invasive fungal infection in these patients is similar to that seen in other clinical settings. As with other transplantation procedures, the vascular anastomosis may become involved and rupture, and there is evidence that the organism in these cases may have been contaminating or infecting the graft on transplantation [34]. Candiduria may indicate the presence of disseminated infection or may progress to the formation of fungal balls (which may obstruct the allograft) or fungal pyelonephritis.

Fungal balls, or bezoars, are chiefly aggregates of fungal hyphae, occasionally containing debris, necrotic renal papillae or even stone material. Infection of the graft itself may be found on renal biopsy. Hyphae or pseudohyphae are typically seen within the tubules and interstitium. Graft infection with *C. albicans* characteristically causes microabscess formation and the fungus may be identified by microscopy and culture of a fine needle aspirate. Graft infection with *Aspergillus* species or the zygomycetes usually shows evidence of invasion of vascular structures and consequent areas of necrosis.

Cryptococcal infection occurs in 2%–4% of these patients [35–37]. It usually presents with fever and meningitis, although occasionally pneumonitis, retinitis, arthritis, skin lesions and pyelonephritis may be seen. Diagnosis may be made on histology (with demonstration of the characteristic xylomannan capsule by mucicarmine red or other special stains) or culture, or by antigen detection, using enzyme-linked immunosorbent assay (ELISA) or latex agglutination methods. Both of these latter techniques are highly specific and achieve sensitivities in excess of 90% in the cerebrospinal fluid.

Infection with the endemic or "pathogenic fungi" are seen during the mid- to late post-transplant periods. Both primary infection (arising from a graft from an infected donor [38], associated with hospital building work in an endemic area [39] or following community exposure post-transplant) and reactivation are seen. Coccidioidomycosis has been reported in 7% of renal transplants in a study in Arizona [40].

Disseminated fungal infection in renal-transplant recipients carries a high mortality (up to 90% with disseminated candidosis and invasive aspergillosis and more than 30% with cryptococcosis). Treatment of invasive fungal infections in these patients is based largely on experience gained from all groups of immunocompromised patients, and there have been very few studies performed specifically in renal-transplant recipients. *Pneumocystis carinii* infections are usually responsive to high-dose co-trimoxazole therapy, but most other invasive fungal infections are treated with i.v. amphotericin B. Alternatives to this nephrotoxic agent include fluconazole (which has been shown to be as effective as amphotericin B in treating candidaemia in non-neutropenic patients [41]) and itraconazole, which, unlike fluconazole, has activity against *Aspergillus* species. Nephrotoxicity may also be overcome by using lipid-based preparations of amphotericin B.

3.4 Parasitic Infections

The main parasitic infections seen in all groups of transplant recipients are those due to *Toxoplasma gondii* and *Strongyloides stercoralis*. However, disease due to these organisms is very uncommon in renal-transplant recipients, with less than 1% developing primary toxoplasmosis in one study [42]. Most of the latter cases occur within 2 months of transplantation and take the form of encephalitis, brain abscess, retinitis, pneumonitis, cardiac involvement and hepatitis [43, 44].

Diagnosis is carried out by means of serology [a significant rise in immunoglobulin M (IgM)] in a patient with a typical ring-enhancing lesion on computed tomography (CT) scan (although antibody production is less reliable in transplant patients), isolation of the organism itself from tissue, or histology. Lesions are characterised by the presence of trophozoites, surrounded by an inflammatory infiltrate in sections stained with Giemsa or periodic acid–Schiff (PAS). The organism may be confirmed by immunohistochemistry and, more recently, PCR has been used, although choice of primers is important as problems have been encountered with earlier studies that amplified a sequence of the P30 gene, subsequently found to occur in a whole host of unrelated organisms.

The treatment of choice for toxoplasmosis remains pyrimethamine together with suphadiazine or clindamycin. *Strongyloides* infection following transplantation is very unusual in the UK, but should be considered in any patient who has travelled to an endemic area. The infection is common worldwide, particularly in the Far East, and also occurs in the southern USA. Pre-transplant work-up should include the screening of stools in such patients, and those found to be infected should be treated prior to transplantation with thiabendazole. Infection during the post-transplant period may take the form of intestinal strongyloidiasis, characterised by abdominal pain, diarrhoea, distension, nausea and vomiting, or hyperinfection syndrome, either confined to the gastrointestinal tract and lungs, or disseminated to other sites, such as the central nervous system, skin and heart.

Disseminated infection carries a mortality of more than 70% [45]. Diagnosis is confirmed by finding larvae in aspirates from affected sites, including the jejunum, or raised titres of antibody. Treatment is with thiabendazole or ivermectin.

4 Liver Transplantation

4.1 Bacterial Infections

Bacterial infection occurs in up to 70% of liver transplant recipients [46–49]. The most common types of infection are: intra-abdominal abscess, cholangitis, bacteraemia, wound infection, lower respiratory-tract infection and urinary-tract infection, with intra-abdominal infection accounting for approximately 30%. Mortality associated with these infections is approximately 4%. Most of these infections are associated with surgical complications and occur within the first 2 months of transplantation. Intra-abdominal sepsis may be the consequence of biliary leak, bile-duct obstruction or hepatic-artery thrombosis (leading to hepatic ischaemia and necrosis). Studies have shown that the technique used to anastomose the donor biliary tract with the recipient has an impact on the incidence of biliary sepsis. Direct duct-to-duct anastomosis carries a much lower risk [50].

Typical organisms isolated from these infected sites include gram-negative aerobes, such as *Escherichia coli*, enterococci, anaerobes and staphylococci. The use of imaging in the diagnosis of these infections is important. Computerised tomography (CT) scanning is valuable in demonstrating collections of intra-abdominal pus and intra-hepatic lesions. Various techniques, including ultrasonography and angiography, may be used to investigate the patency of the hepatic artery. Cholangiography, frequently utilised to determine the integrity of the biliary tract, has the added benefit of allowing samples of bile to be taken for culture.

Liver-transplant patients have a similar risk of opportunistic bacterial infections, such as legionnaire's disease, as renal-transplant recipients.

4.2 Viral Infections

Cytomegalovirus infections have a similar incidence, chronology and presentation as they do in renal-transplant recipients except that hepatitis occurs much more frequently. This typically presents with fever and raised serum alkaline phosphatase, together with elevation of the aminotransferases. Diagnosis and treatment are the same as in renal-transplant patients.

Epstein-Barr virus infections are, again, similar to those seen in the renal-transplant patient, although the incidence of PTLD is higher (approximately 2%) [51–53]. However, the virus may cause severe hepatitis, which may mimic acute rejection.

HSV and VZV infections occur with a similar incidence to those in renal-transplant recipients. On rare occasions, VZV infection has been associated with fulminant hepatic necrosis.

Transplantation for hepatitis B virus-associated liver disease is usually followed by re-infection of the donor liver. Recurrent infection is highest in patients who are HBeAg positive (more than 80% [54]) or HBV DNA positive, both a reflection of high levels of circulating virus. It is much less common in patients who have co-infection with hepatitis D virus (who have much lower levels of circulating virus). Immunoprophylaxis with HBV immunoglobulin (HBIg) has been shown to reduce the recurrence rate, and recent studies with lamivudine given pre- and post-operatively have shown promise [55].

The most common cause of post-transplant hepatitis is hepatitis C virus (HCV). The majority of infections with this agent occur as a result of re-infection in patients who have been transplanted for HCV related cirrhosis. PCR techniques have shown that virtually all infected patients suffer re-infection post-transplant. One study has shown that 95% of such patients developed post-transplant hepatitis and the majority of cases were found to be due to HCV. At 1 year post-transplant, 56% had no histological evidence of chronic liver disease [56].

The impact of hepatitis G virus (HGV) on liver-transplant recipients is still uncertain. Up to 25% of those undergoing transplantation for end-stage HCV liver disease have been found to be co-infected with HGV, but no clinical effects linked to the virus could be detected and co-infection had no effect on incidence or severity of HCV infection post-transplant [57, 58].

Human herpes virus-6 is still of uncertain pathogenicity in transplant recipients, but it has been associated with clinical disease in up to 11% of liver-transplant patients [59]. Clinical features have included myelosuppression, pneumonitis, rash, fever and encephalopathy.

In addition to urinary-tract and respiratory-tract infection adenovirus can cause severe hepatitis in these patients [60, 61].

4.3 Fungal Infections

The incidence of fungal infection varies from 4%–50% in different series [5, 47–49, 62–65]. The majority of these are caused by *Candida* species (77%–83%), with *Aspergillus* species causing most of the rest [5, 48]. *Pneumocystis carinii* pneumonitis occurs in 0%–11% of patients and is closely linked with CMV disease [47–49]. Cryptococcosis has occurred in up to 2% [5]. Cases of zygomycosis [5] and infection with *Pseudallescheria boydii* [66] have also been described, but these seem rare.

More than 50% of fungal infections originate from the abdominal cavity [64] with 15%–20% of cases involving the lungs and fungaemia or disseminated infection occurring in up to 17% [5, 49]. Although very infrequent, aspergillosis is the most common cause of focal brain infection in this group of patients and *Cryptococcus neoformans* the most frequent cause of meningitis.

Severe invasive fungal infections carry a high mortality. Candidal infections have a crude mortality of 50%–77% and invasive aspergillosis is almost universally fatal in this group [67].

The investigation and treatment of fungal infections is as for renal-transplant recipients.

4.4 Parasitic Infections

20% of seronegative recipients receiving a graft from a seropositive donor develop evidence of primary toxoplasmosis [42]. The diagnosis and treatment is as for renal transplant-related infections.

5 Heart Transplantation

5.1 Bacterial Infection

Bacterial infection occurs in up to 30% of heart-transplant recipients. As might be expected, a major site of sepsis is the mediastinum, where the great vessel anastomoses may become infected, causing aneurysm formation or rupture. However, the predominant infections in these patients are in the lower respiratory tract [68]. Others include sternotomy infections, bacteraemia from indwelling cannulae (and rare cases of endocarditis) and urinary-tract infections.

As is the case for renal-transplant recipients, cardiac-transplant patients are susceptible to nocardiosis. Although pulmonary infection is the most common presentation, these individuals are particularly prone to mediastinitis [69].

Involvement of the myocardium of the donor heart may occur in disseminated *Listeria* infection [70]. Other features of this infection are similar to those seen in other transplant groups.

5.2 Viral Infections

Cytomegalovirus infection is the most common viral infection in these patients, with symptomatic disease occurring in up to 25% [16]. Clinical presentation is as for other transplant groups. However, CMV myocarditis is almost unique in this population [71].

Epstein-Barr-related PTLD occurs with a similar frequency to that seen in renal-transplant recipients, but treatment of the condition is rendered difficult by the problems encountered by reducing immunosuppression, with the consequent risk of rejection of this critical organ. Prognosis is therefore worse in this group.

5.3 Fungal Infections

Fungal infections occur with slightly greater frequency than in renal-transplant recipients, but less than in liver-transplant patients. Features of these are similar to other groups, but again the nature of the transplant procedure gives rise to characteristic infections. Thus, *Candida* mediastinitis occurs in this group and candidal infection of the anastomoses of the great vessels may result in sudden death from rupture or aneurysm formation. *Candida* endocarditis is a rare, but frequently fatal condition in these patients and usually requires surgical intervention for cure.

The incidence of invasive aspergillosis is approximately 4% [72]. As in other transplant recipients, pulmonary infection is most common, but cardiac involvement is more common in these patients than in other groups, with pericarditis, myocardial invasion, anastomotic infection and endocarditis proving almost universally fatal. Wound infections may progress to local bone involvement and mediastinitis and should, therefore, be treated aggressively.

5.4 Parasitic Infections

These patients are at greatest risk of toxoplasmosis. More than 50% of seronegative patients receiving a heart from a seronegative donor will acquire primary infection. Of these, approximately 40% will be symptomatic [73]. The infection is readily diagnosed by serological investigations including IgM detection [73].

6 Lung and Heart–Lung Transplantation

6.1 Bacterial Infections

More than 50% of lung-transplant recipients suffer bacterial infections. Patients transplanted for cystic fibrosis are at similar overall risk of infection as those transplanted for other conditions, although they are at increased risk of fatal infection with *Burkholderia cepacia* in the early post-transplant period [74]

Pulmonary infection is the most common bacterial infection in these patients, typically occurring in the first 2 weeks post-transplantation. Causative organisms are usually gram negative, particularly *Klebsiella pneumoniae, Pseudomonas aeruginosa, E. coli* and *Enterobacter cloacae,* although gram-positive infections, such as those due to *Staphylococcus aureus*, are also important [75]. These patients are also at risk of mediastinitis as a consequence of airway anastomotic leaks.

6.2 Viral Infections

Symptomatic CMV infection occurs in approximately 40% and pneumonitis is more common than in other transplant groups [16]; CMV infection has been associated with chronic graft rejection. Lung biopsy may show evidence of obliterative bronchiolitis or mononuclear perivascular infiltration [76, 77].

Episodes of other respiratory viral infections are common in these patients. Causative agents include the paramyxoviruses, respiratory syncytial virus (RSV) and parainfluenza and, as in the general population, they are more common in the under 18 years-of-age group [78]. Patients with these infections may present with cough, dyspnoea, fever, wheezing or coryza. RSV infection may respond to nebulised ribavirin. The incidence of influenza virus infections, on the other hand, appears to be similar to that of the general population.

6.3 Fungal Infections

In different studies, invasive fungal infection has occurred in up to 22% of these transplant recipients [79–81]. Like heart-transplant recipients, heart–lung-transplant patients may suffer infection of great-vessel anastomoses with *Candida* species, and both lung- and heart–lung-transplant recipients are at risk of candidal mediastinitis.

Of all transplant groups, lung-transplant patients are at greatest risk of invasive aspergillosis. Up to 15% are affected. Whilst patients may present with the typical features of invasive pulmonary aspergillosis, they may suffer tracheobronchitis in the region of the airway anastomosis [82].

Pneumocystis carinii causes infection in up to 15% of lung-transplant recipients. The clinical features and management are the same as for other transplant groups.

7 Other Transplant Recipients

Wound and intra-abdominal infections are common in pancreas- and small bowel-transplant recipients and translocation of bacteria with consequent bacteraemia appears to occur frequently in small-bowel transplantation.

CMV pancreatitis, although rare, occurs most commonly in pancreatic-transplant patients. Of all transplant groups, those undergoing small-bowel transplantation are at greatest risk of PTLD. Up to 15% have been found to suffer this condition [53, 83]

Pancreatic-transplant recipients appear at greatest risk of candidal infection. The procedure carries an incidence of up to 30%.

8 Conclusions

The variety and incidence of infections occurring in transplant recipients are likely to continue to evolve as surgical techniques and immunosuppression regimens are refined and prophylactic strategies become more effective. However, the infection risks of some of the more radical procedures only recently being attempted are much greater than for renal transplantation and, unless the need for antirejection therapy is reduced or eliminated, perhaps by the use of genetically engineered grafts, opportunistic infections will remain a predictable complication of the transplantation process.

References

1. Brayman KL, Stephanian E, Matas AJ, Schmidt W, Payne WD, Sutherland DER, Gores PF, Najarian JS, Dunn DL (1992) Analysis of complications occurring after solid organ transplantation. Arch Surg 127: 38–47
2. Wyner LM (1994) The evaluation and management of urinary tract infections in recipients of solid-organ transplants. Semin Urol 12: 134–139
3. Maki DG, Fox BC, Kuntz J, Sollinger HW, Belzer FO (1992) A prospective, randomized, double-blind study of trimethoprim–sulfamethoxazole for prophylaxis of infection in renaltransplantation. Side effects of trimethoprim sulfamethoxalzole, interaction with cyclosporin. J Lab Clin Med 119: 11–24
4. Tolkoff-Rubin NE, Cosimi AB, Russell PS, Rubin RH (1982) A controlled study of trimethoprim–sulfamethoxazole prophylaxis of urinary tract infection in renal transplant recipients. Rev Infect Dis 4: 614–618
5. Wajszczuk CP, Dummer JS, Ho M, Van Thiel DH, Starzl TE, Iwatsuki S, Shaw BJ (1985) Fungal infections in liver transplant recipients. Transplantation 40: 347–353
6. Nelson PW, Delmonicao FL, Tolkoff-Rubin NE, et al. (1984) Unsuspected donor Pseudomonas infection causing arterial disruption after renal transplantation. Transplantation 37: 313–314
7. McCoy GC, Loening S, Braun WE, et al. (1975) The fate of cadaver renal allografts contaminated before transplantation. Transplantation 20: 467–472
8. Majeski JA, Alexander JW, First MR, et al. (1982) Transplantation of microbially contaminated cadaver kidneys. Arch Surg 117: 221–224
9. Fernando ON, Higgins AF, Moorhead JF (1976) Secondary haemorrhage after renal transplantation. Lancet 2: 368
10. Hoepelman IM, Bakker LJ, Jessurun RF, Rozenberg-Arska M, Verhoef J (1987) Disseminated Nocardia asteroides infection complicating renal transplantation. Neth J Med 31: 175–182
11. Rao KV, T O'Brien J, Andersen RC (1987) Septic arthritis due to Nocardia asteroides after successful kidney transplantation. Arthritis Rheum 24: 99–101
12. Sack K, Schwieder G, Marre R, Hoyer J (1985) Nocardial infection in a renal transplant recipient – a case report. Scand J Urol Nephrol Suppl 92: 59–66
13. Huang JY, Huang CC, Lai MK, Chu SH, Chuang CK (1994) Salmonella infection in renal transplant recipients. Transplant Proc 26: 2147
14. Patterson WJ, Hay J, Seal DV, McLuckie JC (1997) Colonization of transplant unit water supplies with *Legionella* and protozoa: precautions required to reduce the risk of legionellosis. J Hosp Infect 37: 7–17
15. Qunibi WY, al-Sibai MB, Taher S, Harder EJ, de Vol E, al-Furayh O, Ginn HE (1990) Mycobacterial infection after renal transplantation – report of 14 cases and review of the literature. QJM 77: 1039–1060
16. Ho M (1994) Advances in understanding cytomegalovirus infection after transplantation. Transplant Proc 26: 7–11
17. Richardson W, Colvin RB, Cheeseman SH, Tolkoff-Rubin NE, Herrin JT, Cosimi AB, Collins AB, Hirsch MS, McCluskey RT, Russell PS, Rubin RH (1981) Glomerulopathy associated with cytomegalovirus viremia in renal allografts. N Engl J Med 305: 57–63
18. Kidd M, Fox JC, Pillay D, Charman H, Griffiths P, Emery V (1993) Provision of prognostic information in immunocompromised patients by routine application of the polymerase chain reaction for cytomegalovirus. Transplant Proc 56: 867–871
19. Patel R, Smith TF, Espy MJ, Wiesner RH, Krom RAF, Portela D, Paya CV (1994) Detection of cytomegalovirus DNA in sera of liver transplant recipients. J Clin Microbiol 32: 1431–1434
20. Spector SA, Merrill R, Wolf D, Dankner WM (1992) Detection of human cytomegalovirus in plasma of AIDS patients during acute visceral disease by DNA amplification. J Clin Microbiol 30: 2359–2365
21. Patel R, Smith TF, Espy M, Portela D, Wiesner RH, Krom RAF, Paya CV (1995) A prospective comparison of molecular diagnostic techniques for the early detection of cytomegalovirus in liver transplant recipients. J Infect Dis 171: 1010–1014

22. Rostaing L, Icart J, Durand D, Henry S, Lloveras JJ, Didier J, Suc JM (1993) Clinical outcome of Epstein-Barr viraemia in transplant patients. Transplant Proc 25: 2286–2287
23. Thomas DW, Ramsahoye B, Jasani B, Lim SH (1995) Epstein-Barr virus in squamous cell carcinoma after renal transplantation. Transplantation 60: 390–392
24. Lee ES, Locker J, Nalesnik M, Reyes J, Jaffe R, Alashari M, Nour B, Dickman PS (1995) The association of Epstein-Barr virus with smooth-muscle tumors occurring after organ transplantation. N Engl J Med 332: 19–25
25. Davis C, Gretch DR, Carithers RL (1995) Hepatitis B and transplantation. Infect Dis Clin North Am 9: 925–941
26. Goffin E, Pirson Y, Cornu C, Guebel A, Squifflet J-P, van Ypersele de Strihou C (1994) Outcome of HCV infection after renal transplantation. Kidney Int 45: 551–555
27. Roth D, Zucker K, Cirocco R, DeMattos A, Burke CW, Nery J, Esquenazi V, Babischkin S, Miller J (1994) The impact of hepatitis C virus infection on renal allograft recipients. Kidney Int 45: 238–244
28. Terrault NA, Wright TL, Pereira BJG (1995) Hepatitis C infection in the transplant recipient. Infect Dis Clin North Am: 943–964
29. Rubin RH (1994) Infection in the organ transplant recipient. In: Rubin RH, Young LS (eds) Clinical approach to infection in the compromised host, 3rd edn. Plenum, New York, pp 629–705
30. Ogunbiyi OA, Scholefield JH, Raftery AT, Smith JHF, Duffy S, Sharp F, Rogers K (1994) Prevalence of anal human papillomavirus infection and intraepithelial neolplasia in renal allograft recipients. Br J Surg 81: 365–367
31. Leapman SB, Rosenberg JB, Filo RS, Smith EJ (1980) *Strongyloides stercoralis* in chronic renal failure: safe therapy with thiabendazole. South Med J 73: 1400–1402
32. Blohme I, Nyberg G, Jeansson S, Svalander S (1992) Adenovirus infection in a renal transplant patient. Transplant Proc 24: 295
33. Denning DW, Evans EGV, Kibbler CC, Richardson MD, Roberts MM, Rogers TR, Warnock DW, Warren RE (1997) Guidelines for the investigation of invasive fungal infections in haematological malignancy and solid organ transplantation. Eur J Clin Microbiol Infect Dis 16: 424–436
34. McLeish KR, McMurray SD, Smith EJ, Filo RS (1977) The transmission of Candida albicans by cadaveric allografts. J Urol 118: 513–516
35. Kong NCT, Suleiman AB, Shaarian W, Wong YH, Morad Z (1990) Cryptococcosis in a renal unit. Aust N Z J Med 20: 645–679
36. Shaariah W, Morad Z, Suleiman AB (1992) Cryptococcosis in renal transplant recipients. Transplant Proc 24: 1898–1899
37. Watson AJ, Russell RP, Cabreja RF, Braverman R, Whelton A (1985) Cure of cryptococcal infection during continued immunosuppressive therapy. QJM 55: 169–172
38. Gottesdiener KM (1989) Transplanted infections: donor-to-host transmission with the allograft. Ann Intern Med 110: 1001–1016
39. Wheat LJ, Smith EJ, Sathapatayavongs B, Batteiger B, Filo RS, Leapman SB, French MV (1983) Histoplasmosis in renal allograft recipients. Two large urban outbreaks. Arch Intern Med 143: 703–707
40. Cohen IM, Galgiani JN, Potter D, Ogden DA (1982) Coccidioidomycosis in renal replacement therapy. Arch Intern Med 142: 489–494
41. Rex JH, Bennett JE, Sugar AM, et al. for the Candidemia Study Group and the National Institute of Allergy and Infectious Diseases Mycoses Study Group (1994) A randomized trial comparing fluconazole with amphotericin B for the treatment of candidemia in patients without neutropenia. N Engl J Med 331: 1325–1330
42. Speirs GE, Hakim M, Wreghitt TG (1988) Relative risk of donor transmitted Toxoplasma gondii infection in heart, liver and kidney transplant recipients. Clin Transplant 2: 257–269
43. Michaels MG, Wald ER, Fricker FJ, del Nido PJ, Armitage J (1992) Toxoplasmosis in pediatric recipients of heart transplant. Clin Infect Dis 14: 847–851
44. Singer MA, Hagler WS, Grossniklaus HE (1993) Toxoplasma gondii retinochoroiditis after liver transplantation. Retina 13: 40–45

45. DeVault GA, King JW, Rohr MS, Landreneau MD, Brown ST, McDonald JC (1990) Opportunistic infections with Strongyloides stercoralis in renal transplantation. Rev Infect Dis 12: 653–671
46. George DL, Arnow PM, Fox AS, Baker AL, Thistlethwaite JR, Emond JC, Whitington PF, Broelsch CE (1991) Bacterial infection as a complication of liver transplantation: epidemiology and risk factors. Rev Infect Dis 13: 387–396
47. Paya CV, Hermans PE, Washington JA, Smith TF, Anhalt JP, Wiesner RH, Krom RA (1989) Incidence, distribution, and outcome of episodes of infection in 100 orthotopic liver transplantations. Mayo Clin Proc 64: 555–564
48. Kusne S, Dummer JS, Singh N, Iwatsuki S, Makowka L, Esquivel C, Tzakis AG, Starzl TE, Ho M (1988) Infections after liver transplantation. An analysis of 101 consecutive cases. Medicine (Baltimore) 67: 132–143
49. Colonna JO, Winston DJ, Brill JE, Goldstein LI, Hoff MP, Hiatt JR, Quinones Baldrich W, Ramming KP, Busuttil RW (1988) Infectious complications in liver transplantation. Arch Surg 123: 360–364
50. Kibbler CC (1995) Infections in liver transplantation: risk factors and strategies for prevention. J Hosp Infect 30 (Suppl): 209–217
51. Garnier JL, Berger F, Betuel H, Vuillaume M, Chapuis-Cellier C, Blanc N, Faure JL, Dubernard JM, Lenoir G, Touraine JL (1989) Epstein-Barr virus associated lymphoproliferative diseases (B cell lymphoma) after transplantation. Nephrol Dial Transplant 4: 818–823
52. Nalesnik MA, Jaffe R, Starzl TE, Demetris AJ, Porter K, Burnham JA, Makowka L, Ho M, Locker J (1988) The pathology of posttransplant lymphoproliferative disorders occurring in the setting of cyclosporine A–prednisone immunosuppression. Am J Pathol 133: 173–192
53. Walker RC, Paya CV, Marshall WF, Strickler JG, Wiesner RH, Velosa JA, Habermann TM, Daly RC, McGregor CGA (1995) Pretransplantation seronegative Epstein-Barr virus status is the primary risk factor for posttransplantation lymphoproliferativc disorder in adult heart, lung, and other solid organ transplantations. J Heart Lung Transplant 14: 214–221
54. Samuel D, Muller R, Alexander G (1993) Liver transplantation in European patients with the hepatitis B surface antigen. N Engl J Med 329: 1842–1847
55. Grellier L, Mutiner D, Ahmed M, Brown D, Burroughs AK, Rolles K, McMaster P, Beranek P, Kennedy F, Kibbler H, et al. (1996) Lamivudine prophylaxis against reinfection in liver transplantation for hepatitis B cirrhosis. Lancet 348: 1212–1215
56. Wright TL, Donegan E, Hsu HH, Ferrell L, Lake JR, Kim M, Combs C, Fennessy S, Roberts JP, Ascher NJ (1992) Recurrent and acquired hepatitis C viral infection in liver transplant recipients. Gastroenterol 103: 317–322
57. Brandhagen DJ, Gross JB, Poterucha JJ, Kim WR, Charlton WR (1996) Hepatitis G infection as determined by bDNA in patients with hepatitis C undergoing liver transplantation: clinical characteristics and viral levels (abstract). Hepatology 24: 229A
58. Poutous A, Vargas H, Laskus T, Wang L, Radkowski M (1996) Hepatitis G infection in HCV-positive liver transplant recipients (abstract). Hepatology 24: 420A
59. Singh N, Carrigan DR (1996) Human herpesvirus-6 in liver transplantation: documentation of pathogenicity (abstract). 36th Interscience Conference on Antimicrobial Agents and Chemotherapy, New Orleans, H12
60. Salt A, Sutehall G, Sargaison M, Woodward C, Barnes ND, Calne RY, Wreghett TG (1990) Viral and Toxoplasma gondii infections in children after liver transplantation. J Clin Pathol 43: 63–67
61. Zitelli BJ, Gartner JC, Laltack JJ, Urbach AH, Miller JW, Williams L, Kirkpatrick B, Breinig MK, Ho M (1987) Pediatric liver transplantation: patient evaluation and selection, infectious complications, and life-style after transplantation. Transplant Proc 19: 3309–3316
62. Kirby RM, McMaster P, Clements D, Hubscher SG, Angrisani L, Sealey M, Gunson BK, Salt PJ, Buckels JA, Adams DH, et al. (1987) Orthotopic liver transplantation: postoperative complications and their management. Br J Surg 74: 3–11
63. Castaldo P, Stratta RJ, Wood RP, Markin RS, Patil KD, Shaefer MS, Langnas AN, Shaw BW (1991) Fungal infections in liver allograft recipients. Transplant Proc 23: 1967
64. Castaldo P, Stratta RJ, Wood RP, Markin RS, Patil KD, Shaefer MS, Langnas AN, Reed EC, Li SJ, Pillen TJ, et al. (1991) Clinical spectrum of fungal infections after orthotopic liver transplantation. Arch Surg 126: 149–156

65. Viviani MA, Tortorano AM, Malaspina C, Colledan M, Paone G, Rossi G, Bordone G, Pagano A (1992) Surveillance and treatment of liver transplant recipients for candidiasis and aspergillosis. Eur J Epidemiol 8: 433–436
66. Patterson TF, Andriole VT, Zervos MJ, Therasse D, Kauffman CA (1990) The epidemiology of pseudallescheriasis complicating transplantation: nosocomial and community-acquired infection. Mycoses 33: 297–302
67. Paya CV (1993) Fungal infections in solid-organ transplantation. Clin Infect Dis 16: 677–688
68. Stinson EB, Bieber CP, Griepp RB, Clark DA, Shumway NE, Remington JS (1971) Infectious complications after cardiac transplantation in man. Ann Intern Med 74: 24–36
69. Thaler F, Gotainer B, Teodori G, Dubois C, Loirat P (1992) Mediastinitis due to Nocardia asteroides after cardiac transplantation. Intensive Care Med 18: 127–128
70. Stamm AM, Smith S-H, Kirklin JK, McGiffin DC (1990) Listerial myocarditis in cardiac transplantation. Rev Infect Dis 12: 820–823
71. Grossi P, Revello G, Minoli L, Percivalle E, Zavattoni M, Poma G, Martinelli L, Gerna G (1990) Three-year experience with human cytomegalovirus infections in heart transplant recipients. J Heart Transplant 9: 712–719
72. Guillemain R, Lavarde V, Amrein C, Chevalier P, Guinvarc'h A, Glotz D (1995) Invasive aspergillosis after transplantation. Transplant Proc 27: 1307–1309
73. Gallino A, Maggiorini M, Kiowski W (1996) Toxoplasmosis in heart transplant recipients. Eur J Clin Microbiol Infect Dis 15: 389–393
74. Ramirez JC, Patterson GA, Winton T, de Hoyos AL, Miller JD, Maurer JR, and The Toronto Lung Transplant Group (1992) Bilateral lung transplantation for cystic fibrosis. J Thorac Cardiovasc Surg 103: 287–294
75. Deusch E, End A, Grimm M, Graniger W, Klepetko W, Wolner E (1993) Early bacterial infections in lung transplant recipients. Chest 104: 1412–1416
76. Allen M, Burke CM, McGregor CG, Baldwin JC, Jamieson SW, Theodore J (1986) Steroid responsive bronchiolitis after human heart lung transplantation. J Thorac Cardiovasc Surg 92: 449–451
77. Maurer JR, Tullis DE, Scavuzzo M, Patterson BS, Patterson GA (1991) Cytomegalovirus infection in isolated lung transplantations. J Heart Lung Transplant 10: 647–649
78. Wendt CH, Fox JMK, Hertz Ml (1995) Paramyxovirus infection in lung transplant recipients. J Heart Lung Transplant 14: 479–485
79. Brooks RG, Hofflin JM, Jamieson SW, Stinson EB, Remington JS (1985) Infectious complications in heart-lung transplant recipients, Am J Med 79: 412–422
80. Dummer JSC, Montero CG, Griffith BP, Hardesty RL, Paradis L, Ho M (1986) Infections in heart-lung transplant recipients. Transplantation 41: 725–729
81. Hofflin JM, Potasman I, Baldwin JC, Oyer PE, Stinson EB, Remington JS (1987) Infectious complications in heart transplant recipients receiving cyclosporine and corticosteroids. Ann Intern Med 106: 209–216
82. Kramer MR, Denning DW, Marshall SE, Ross DJ, Berry G, Lewiston NJ, Stevens DA, Theodore J (1991) Ulcerative tracheobronchitis after lung transplantation. A new form of invasive aspergillosis. Am Rev Respir Dis 144: 552–556
83. White FV, Reyes J, Jaffe R, Yunis EJ (1995) Pathology of intestinal transplantation in children. Am J Surg Pathol 19: 687–698

Chronic Allograft Nephropathy: The Inevitable Outcome of Renal Transplantation?

S. M. Dodd

1 Introduction

Many renal allografts now survive and function for decades, but long-term degenerative remodelling of the transplant (chronic allograft nephropathy), with inexorable decline in renal function, is one of the most important determinants of the ultimate outcome of renal transplantation [86].

Two phases of graft loss are recognised; in the first phase, which is defined as the early postoperative period (up to 100 days), between 5% and 20% of grafted kidneys cease to function [15]. Success at this early stage is determined by: (1) the skills of individual transplant centres in avoiding non-immunological causes of graft loss, (2) matching the donor and recipient at human leucocyte antigen (HLA) foci, and (3) applying appropriate immunosuppression for episodes of acute

Current Topics in Pathology
Volume 92, C. L. Berry (Ed.)

immunologically mediated rejection. At these centres, 1-year survival rates for cadaveric transplants can be greater than 90%, indicating that the problems associated with acute rejection have largely been overcome. Cyclosporin immunosuppressive therapy has been most effective during this period [87].

Although increased short-term survival contributes to improved long-term success, the annual rate of graft loss after the first year has not changed significantly [15]. It is appropriate to distinguish between patient and graft loss in this later phase. Over the first 10 years, approximately one-third of patients die with a functioning graft mainly due to cardiovascular disease (a particular problem in diabetic patients), infection and malignancy. The remaining two-thirds of all transplants succumb to acute or chronic immunologically mediated rejection during this period.

2 Definition of Chronic Rejection

Chronic rejection was defined in 1993 during a consensus conference as a progressive deterioration of graft function (in at least two measurements at a 3-month interval beginning at least 3 months after transplantation) in the absence of any other disorder and after confirmation of the diagnosis by pathological study [104]. To this definition may be added the observation that the lesion does not improve following administration of drugs that are usually effective in reversing acute rejection episodes [63]. Its pathophysiology and aetiology are incompletely understood and, in its normal usage, the term chronic rejection describes a chronic, pathological situation that results from an association of multiple immunological processes that may or may not be linked to the allograft response with non-immunological processes that may have pre-existed in the donor, been induced by the transfer of the organ or developed subsequently in the grafted organ.

The clinical manifestations of chronic rejection consist of proteinuria, a progressive decrease in glomerular filtration rate and hypertension [102, 143]. It is certain that the pathological onset of chronic rejection happens much earlier than its clinical onset, as demonstrated by the high percentage of pathological lesions compatible with those of chronic rejection seen in grafts with normal function.

3 Histopathology of Chronic Allograft Nephropathy

The Banff working classification of renal transplant pathology [125, 126] developed a schema for international standardisation of nomenclature and criteria for the histological diagnosis of renal allograft rejection. Four types of rejection were recognised: hyperacute rejection, borderline changes (very mild acute rejection), acute rejection and chronic allograft nephropathy (chronic rejection). Hyperacute rejection, seen extremely rarely in clinical practise, is presumed to be due to the presence of pre-formed antibody, and is characterised by polymorph accumulation in glomerular and peritubular capillaries at 1 h post-transplant, with subsequent endothelial damage and capillary thrombosis.

The histopathological criteria used to diagnose acute rejection in this classification have a different emphasis, with tubulitis and intimal arteritis considered to be the principal lesions [20, 88, 122], whereas, in the past, the degree of interstitial inflammation was considered to be of paramount importance. Five different studies in which stable kidney transplants were biopsied showed that focal or mild diffuse infiltrates occurred commonly in well-functioning grafts [13, 21, 78, 92, 125, 126]. Chronic allograft nephropathy is characterised in the Banff classification by both chronic transplant glomerulopathy, a lesion in which there is mesangial cell proliferation, mesangial matrix increase and peripheral mesangial interposition with thickening of glomerular capillary loops and also by interstitial fibrosis and tubular atrophy, with proliferation and thickening of the vascular intima. Interstitial inflammation often persists and may be of varying degree. It is now clearly recognised that the tubulointerstitial changes are of a secondary nature and are consequent upon the proliferative endarteritis and concentric intimal fibrosis common to all grafted organs [20, 95, 115]. However, histopathological study of chronic allograft nephropathy cannot distinguish lesions secondary to the immune response induced by the allograft from the many non-immunological factors that lead to graft destruction. Both forms of graft damage will be discussed below.

4 Pathological Processes Responsible for Chronic Allograft Nephropathy

Long-term graft destruction may, in part, be due to an immunological process with its associated inflammatory changes. It is not certain that the mechanisms of acute rejection differ from those governing chronic rejection, except in intensity. However, numerous non-immunological risk factors are also present, such as the condition of the donor organ and various constitutional factors in the recipient. Overlap factors exist, such as infections, which could be considered in either the immunological or non-immunological group of processes.

5 Immunological Factors Promoting Chronic Allograft Destruction

5.1 Persistence of an Immune Activation Process

Although not formally demonstrated in the renal graft recipient, the persistence of a permanent immunological process activating the alloantigen recognition system of the host is likely. One of the most compelling pieces of evidence is the finding that vascular-wall atherosclerosis occurs more rapidly and diffusely in allografts than in isografts or syngeneic grafts. In man, recipients of HLA-identical grafts have a much better survival rate than those receiving cadaveric grafts [98]. The intensity of this could vary, depending on both the major and/or minor histocompatibility-system differences between the graft and the host, and on the nature and intensity of immunosupression and would lead to a more or less rapid destruction of the graft. A very low intensity immune activation, leading to destruction over several decades, would be the equivalent of long-term graft acceptance, a more encouraging way of expressing the concept of chronic rejection of very low activity.

5.2 Antigen Presentation and Recognition

Alloresponsiveness between genetically unrelated humans is directed against a single cluster of antigens, designated HLA and encoded by major histocompatibility complex (MHC) genes found on chromosome 6. The MHC antigen groups, class I (HLA-A and -B) and class II (HLA-D and -DR), which have different tissue distributions, are important targets for host immunoreactivity. A variety of cell mediators or lymphokines regulate the expression of MHC antigens, for example, interferon γ (IFN-γ) upregulates class-II expression on vascular endothelium, renal-tubular epithelial cells and lymphocytes, leading to enhanced antigen presentation and amplification of graft immunogenicity. Interaction between the antigenic receptor of specific T cells to an antigenic peptide bound in a groove of the MHC molecule is required for recognition of the allogeneic peptide as foreign.

Evidence suggests that MHC identity protects against long-term graft rejection [4, 17, 138, 142]. Terasaki's group [139] suggested that the estimated half-life of kidneys transplanted from HLA-matched sibling donors is 20 years. Takemoto's work (1992) showed that the allocation of kidneys according to the current serological methods of HLA typing [67] was sufficient to obtain a 1-year survival rate of 88% and an estimated half-life of 17.3 years. This was significantly higher than the rate of 79% for mismatched transplants with a half-life of 7.8 years in the same collaborative study. However, no study has shown a relationship between HLA typing and histological lesions of chronic rejection [57, 64, 69] and it is clear that many other factors intervene and obscure the MHC effect.

In contrast, recent work from the Netherlands [25] has shown that many allografts function well, even in the presence of one or more HLA mismatches. In the records of Eurotransplant, a multinational, European organ-exchange organisation [145], about 22% of patients have undergone transplantation with a kidney identical or compatible for HLA-A, -B, and -DR (0 HLA mismatches). Doxiadis et al. [25] examined whether the immunogenicity of the different donor HLA mismatches depends on the HLA phenotype of the recipient and found that certain taboo combinations were associated with higher graft loss than others, and that the indifferent mismatched donor–recipient antigen combinations were associated with graft survival almost identical to that of the grafts with no antigen mismatches. The way in which taboo combinations were defined differed from that of previous studies of permissible and immunogenic mismatches [79] in that analysis was restricted to specific donor mismatches and the complete observation time was taken into account, rather than restricting the analysis to a single point in time. Doxiadis et al. [25] found that renal allograft survival was 18% lower with taboo than with indifferent combinations. Taboo donor/recipient HLA combinations occurred in about 30% of the selected single mismatched transplants or 15% of the whole group of transplants with one HLA mismatch in the Eurotransplant database. Their conclusion was that if taboo combinations are avoided in cadaveric renal transplantation, a survival gain of about 3% at 5 years could be achieved.

Interestingly, Terasaki et al. [140] showed that in the United States, where increasing numbers of people are donating kidneys to their spouses, survival rates are higher than those of cadaveric kidneys, despite greater histoincompatibility. They suggested that about 10% of cadaveric grafts are damaged before removal and that this is reflected by a 10% increase in living, unrelated graft survival compared with cadaveric grafts. They also commented that compliance with respect to taking immunosuppressive drugs may be higher among recipients of spousal grafts because the recipient lives with the donor.

5.3 T Cell Recognition of Transplanted Tissues

The acute rejection process is primarily a T lymphocyte-mediated host event and cellular infiltration of the graft is a characteristic feature of acute rejection. Within a few hours of grafting, neutrophils enter the interstitium after interaction with selectins upregulated on vascular cells after the ischaemic operative insult. These cells produce biochemical mediators that further injure and/or activate endothelial cells. Increased expression of cell-adhesion molecules and enhanced vascular permeability result in infiltration of the graft by host lymphocytes and macrophages. Although the graft is infiltrated by both CD4+ and CD8+ cells, it is the CD4+ helper T (Th) cells of the recipient that are central to the initiation of the rejection process and its maintenance. At least two distinct, but not mutually exclusive, pathways of allo-recognition have been recognised. The helper cells may be activated directly by the MHC class II-positive antigen-presenting cells (APCs) of the donor, which are present as passenger leucocytes within the transplant and then migrate into draining lymph nodes [3]. Alternatively, the Th may be activated indirectly by recipient APCs within lymph nodes that take up and process donor antigens shed from the graft. The latter is the normal mechanism by which any foreign protein (such as bacteria or viruses) activates host T cells, whereas only alloantigens have the ability to stimulate T cells directly. Acute T cell-mediated rejection results from the direct activation of Th cells [10], whereas the chronic phase is a consequence of indirect immune activation.

5.4 Mechanisms of Acute Rejection

The function of Th cells, whether activated directly or indirectly, is to produce the lymphokines that are responsible for a range of effector mechanisms that act on the renal allograft to alter its susceptibility to the various processes of rejection [53]. These mechanisms include cytotoxic CD8+ T lymphocytes, antibodies produced by B cells and macrophage-induced inflammatory processes (Fig. 1).

CD4+ cells have been divided into Thl and Th2 sub-populations. Thl cells produce the pro-inflammatory lymphokines, IFN-γ, interleukin 2 (IL-2) and tumour necrosis factor β (TNF-β) and, therefore, activate cell-mediated mechanisms of acute rejection. Th2 cells, however, produce the inhibitory cytokines IL-4, IL-5 and IL-10, which are involved in the activation of B lymphocytes, thereby initiating the

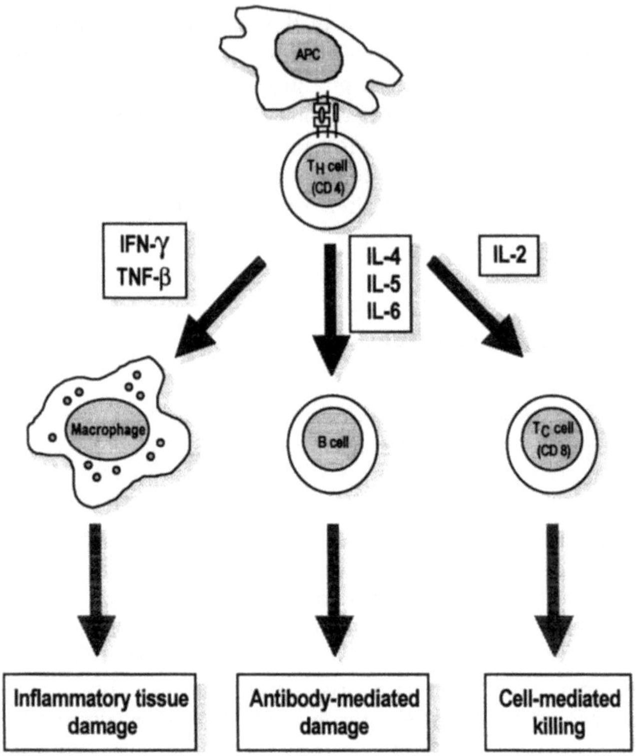

Fig. 1. The role of the T helper cell in graft rejection. *APC,* antigen-presenting cell; *Th cell,* T helper lymphocyte; *IFN,* interferon; *TNF,* tumour necrosis factor; *IL,* interleukin; *B cell,* B lymphocyte; *T cell,* cytotoxic T lymphocyte

humoral mechanisms of acute rejection. Th2 cells also produce transforming growth factor β (TGF-β) and granulocyte–macrophage colony-stimulating growth factor (GM–CSF), both of which play an important role in the activation of T-suppressor cells and in the development of chronic rejection.

CD8+ cells bear a ligand for MHC class-I antigen, which is expressed on most cell surfaces. Two subsets of the CD8+ cell have also been described recently [60]. Type 1 are cytotoxic, produce IFN-γ, but not IL-4, and are restricted to MHC class I. Type II are restricted to MHC class-II antigens, may have a role in suppressing the killing of infected cells, and produce IFN-γ, IL-4, IL-5 and IL-10.

5.5 Regulation of Cytokine Gene Expression

The activation of cytokine genes is dependent on the binding of proteins (so-called transcription factors) to the DNA flanking the genes. The binding of protein factors alters the DNA conformation and allows the access and assembly of the complex of

RNA-polymerase enzymes, required to make an RNA template of the gene. The transcription factors are activated as a final event in the different second messenger pathways that exist in T cells, enabling them to respond differentially to stimuli that selectively activate second-messenger pathways. Some of the protein factors exist in the cytoplasm and, after they are modified by phosphorylation or dephosphorylation, translocate to the nucleus where they bind to DNA. One such factor is nuclear factor of activation of T cells (NF-AT), which is important for the regulation of IL-2, TNF, IFN-γ, IL-4 and others. Cyclosporin and FK506 act to prevent translocation of the cytoplasmic NF-AT to the nucleus, thereby suppressing the activation of cytokine gene expression [123]. Thus, modulation of immunosuppressive agents can be used to alter second-messenger pathways and, hence, the production of cytokines by T cells and certain drugs or combinations of drugs can selectively inhibit the production of targeted cytokines.

5.6 The Role of B Cells and Macrophages in Acute Rejection

Allostimulated B lymphocytes differentiate into antibody-producing plasma cells, which secrete both specific and non-specific anti-donor antibodies [37]. Macrophages act as APCs which activate T cells as well as recruit other macrophages [91]. They secrete monokines, such as IL-1, which stimulates CD4+ lymphocytes to produce other effector cytokines [28]. Natural killer cells are cytotoxic cells that do not express the T-cell antigen, CD3, and have the ability to destroy foreign antigens, independent of the MHC antigen. They are thought to contribute to acute rejection-mediated damage by adhering to the exposed R portions of antibody molecules during antigen–antibody reactions [84]. Natural killer cells may kill allogeneic cells selectively after recognising MHC class-I antigen on their surfaces.

6 Long-Term Graft Survival

In some animal models, the graft recipient can be treated with drugs, antibodies or a variety of antigens at around the time of transplantation, enabling withdrawal of all further immunosuppressive treatment [52]. Unfortunately, this has not been achieved in man, and research efforts have focussed on the possible mechanisms of reducing the immunogenicity of the graft.

6.1 The Induction of Long-Term Graft Acceptance

Experimentally, long-term acceptance of a renal allograft in the rat can be induced by pretreatment with donor antigen or perioperative treatment with anti-donor antibody. Graft acceptance can be achieved by irradiation, with drugs such as cyclosporin or FK506, with antibodies such as anti-lymphocyte globulin or monoclonal antibodies to CD4, IL-2 receptors and T-cell receptors [52]. Certain strains of rats require less immunosuppression to achieve graft acceptance and are considered to be low responders, whereas high responders require more immunosuppression. In man, high and low responders are less easy to define. Some individuals have more rejection episodes and episodes of greater severity, requiring a higher dose of steroids in the early postoperative period; these may be considered high responders. It has also been suggested that recipients with certain haplotypes associated with autoimmunity, such as HLA-Al, -B8 and -DR3, may mount more vigorous rejection responses and, hence, be considered high responders [61]. Low-responder individuals might include the 20% of individuals who failed to become sensitised in the era of deliberate blood transfusion as a means of testing by pre-immunisation; these patients have been shown to have a favourable outcome after 1 year [96].

6.2 Mechanisms of Specific Unresponsiveness

Bone marrow-derived interstitial dendritic cells within tissues provide a powerful stimulus for primary T-cell activation. These passenger leucocytes are not a fixed part of the graft and migrate out of the transplanted organ during the first few days after transplantation to the para-cortical areas of the spleen or lymph nodes and are, here, responsible for direct activation of the immune response. Organs depleted of these passenger leucocytes are far less immunogenic, and long-term survival of such grafts is easier to achieve [68]. This forms the basis for the suggestion that if leucocytes from human donor kidneys are depleted using a monoclonal pan-leucocyte antibody, direct activation of the rejection process will be reduced [12]. Goldberg et al. [39] have shown a reduction in the incidence of acute rejection when human donor kidneys are pretreated with this antibody.

Starzi et al. [130] postulated that the acceptance of whole-organ transplants implies a state of long-term mixed allogeneic or xenogeneic chimerism after the migration of dendritic and other immune cells from the graft with widespread seeding in the recipient; simultaneous chimerism of the graft is produced by a reverse traffic of similar recipient cells. They further suggested [131] that the bi-directional cell migration and re-population is the first step in the acquisition of donor-specific immunological non-responsiveness (tolerance). They demonstrated low-level chimerism by immunocytochemistry and/or polymerase chain reaction in renal allografts, skin, lymph node and blood in five patients, who received continuously functioning renal transplants from one or two haplotype HLA-mismatched, consanguineous donors 27 years ago. The critical migratory cells may be dendritic cells [134], but little information is available as to how these cells of bone marrow origin, whose life expectancy after maturation is only a few days [133], have disseminated from the grafts and perpetuated themselves for nearly 30 years in their new environment. Starzi et al. [130] suggested that drug-free tolerance and graft acceptance achieved with chronic immunosuppression are both expressions of the cell re-population process.

The key role of the Th lymphocyte in graft rejection has been noted above and the two subsets discussed. It is possible that the failure of a recipient to reject a transplant may reflect an inappropriate immune response rather than no response at all. The Th1 cell is important in cell-mediated immunity and the Th2 cell in the antibody response; hence, in acute graft rejection, the Thl subset has a greater role. In rodent models, it has been shown that experimental manoeuvres resulting in immune deviation, away from cell-mediated immunity and towards preferential activation of Th2 cells, is associated with good graft outcome [72]. Lowry also proposed that the selective action of Th2 cells, which produce IL-10, a lymphokine with suppressive effects on inflammatory cells, may be beneficial in terms of graft acceptance.

Joseph et al. [56] tested the importance of selective Th2 action in six human renal-transplant patients, assessing cytokines (TNF-α, IFN-γ, IL-1, IL-2, IL-4, IL-5, IL-6 and IL-10) in serum by enzyme-linked immunosorbent assay (ELISA) and in grafted tissues by reverse transcriptase polymerase chain reaction (RT-PCR) for messenger RNA (mRNA). Localisation of cytokine mRNA in the transplanted tissues was examined using in situ hybridisation. However, the findings were at variance with the animal-model results, with none of the patients making a significant IFN-γ response and only a few producing IL-2. The mRNA for the immunosuppressive cytokine IL-10 was found in areas of active graft damage and repair, and appeared to be a good marker of macrophage activation rather than an indicator of graft function. The study concluded that no single cytokine or consistent combination of cytokines could be correlated with rejection episodes or with graft function. Further work is clearly required before immune deviation by targeting particular T-cell populations can be of therapeutic value in the induction of graft acceptance.

6.3 The Importance of Acute Rejection Episodes in the Evolution of Chronic Allograft Nephropathy

A history of acute rejection is one factor that is clearly related to an increased frequency of chronic rejection in multiple studies. In a study by Basadonna et al. [5], 424 patients alive 1 year after transplantation were examined. Approximately 45% of patients had one or more acute rejection episodes and histopathological assessment of renal biopsy material for features of chronic rejection was made. Only 1 of 239 patients with no prior acute rejection developed biopsy-proven chronic rejection, whereas in the patient group with one episode of acute rejection, 22 of 105 patients developed chronic rejection. There were 80 patients with more than one episode of acute rejection that developed chronic disease. Acute rejection within the first 60 days and late acute rejection were both strongly associated with chronic rejection [2]. Neither of these studies showed that the observed graft destruction is the consequence of an active immunological process. The process could also be that of progression of the sequelae induced by acute rejection. The more frequent, severe and long-lasting these episodes have been, the more severe and irreversible the renal lesions. Furthermore, when a vascular element is seen histologically during the acute rejection episode, progressive degradation of the graft occurs due to ischaemia of the area supplied by the damaged vessels.

Endothelial damage induced by rejection, which is more severe if the recipient was pre-immunised against the donor lymphocytes [19], appears to induce surface expression of adhesion molecules (integrins and selectins), favouring platelet aggregation and adhesion [18]. Subsequent platelet adhesion leads to microthromboses and a reduction in arteriolar diameter, increasing ischaemia. The tubular damage induced by ischaemia would lead to the release of numerous growth factors [8]. Inflammatory cells would then be attracted into the interstitium and react with tubular cells and interstitial fibroblasts [89]. Subsequent fibroblast and perivascular myofibroblast activation would occur with tubulointerstitial fibrosis ensuing.

Each acute rejection episode would be responsible for the destruction of a certain number of nephrons and, ultimately, damage from hyperfiltration would occur [11].

6.4 Immune Mechanisms of Chronic Rejection

Although their existence is highly probable, the immunological mechanisms that would be responsible for chronic rejection are still unknown. Little qualitative difference is seen between the type of cells infiltrating grafts during acute and chronic rejection or even in biopsies performed when the patient has no symptoms of rejection [14, 77, 137]. In the chronic phase of rejection, the HLA match may have less impact on the rate of graft loss [86].

The major histopathological lesion of chronic rejection, from which the others ensue, is vascular damage in the form of a fibroproliferative endarteritis [121]. CD4+ lymphocytes and activated macrophages play a role in the initiation and amplification of the vessel-wall response, and macrophage products, TNF-α, IL-lb, TGF-β and platelet-derived growth factor (PDGF) can be detected. Endothelial cells within the graft are activated with upregulation of intercellular adhesion molecule 1 (ICAM-1) and MHC class-II molecules. Endothelial-cell activation probably results from release of TNF and IFN-γ from inflammatory cells [99], but also occurs due to viral infections and from an antibody effect. Activated endothelial cells can produce cytokines, most importantly TGF-β, which, itself, promotes further release of TGF-β from macrophages and endothelial cells. Both PDGF and TGF- 40 are potent smooth-muscle mitogens, and their action could account for smooth muscle-cell proliferation in the intima, whereas TGF- has powerful fibrogenic properties, accounting for the fibrotic element of the vascular lesions.

7 Non-immunological Lesions Promoting Chronic Allograft Destruction

7.1 Loss of Organ Mass: the Role of Hyperfiltration

Occupying a key position in the evolution of chronic allograft destruction is the loss of organ mass, which may follow a variety of insults, some immunological, and from which ensues an array of haemodynamic and metabolic adaptations that result in or contribute to further damage to the remaining tissue. Surgical ablation of renal tissue produces a variety of changes in the remaining tissue that ultimately lead to damage to the rest of the renal mass [11]. Subsequent to this work, there has been much debate over whether this occurs in man, but a follow-up study of patients who had undergone partial removal of a solitary kidney confirm that similar mechanisms may be operative in man [93]. Brenner's group proposed that it was the sustained increases in glomerular blood flow and pressure in the remaining nephrons which led to structural damage in the remaining glomeruli.

Later studies suggested that the primary determinant leading to structural damage is increased pressure rather than flow. This hypothesis has been challenged and it has been proposed that compensatory glomerular hypertrophy [148], the nature of the increased glomerular permeability [112], the increased ammonia load of the remaining tissue [90] and disorders of lipoprotein metabolism accompanying proteinuria or renal failure [58] also contribute to the structural damage. Modena et al. [83] studied renal transplant patients with chronic rejection and found that the rate of decline in function correlated with the presence of diastolic hypertension, suggesting a role for haemodynamic factors in the progressive loss of function. Experimental studies of chronic rejection in porcine and rat models have demonstrated raised glomerular pressure, glomerular hypertrophy, proteinuria and focal segmental glomerulosclerosis, features characteristic of haemodynamically induced renal injury [48, 62]. The above evidence suggests that a kidney damaged prior to or following renal transplantation by immunological or non-immunological mechanisms will, eventually, on losing a critical proportion of its mass, be subject to progressive glomerular damage and continued tubulointerstitial scarring, irrespective of whether the original insult continues.

7.2 Sufficient Nephron Dosing in the Transplant

Terasaki's group (1994) further applied the hyperfiltration hypothesis to human kidney transplantation and studied five situations in which a real or relative reduction in renal mass is seen: (1) small kidneys from donors aged 4–6 years, (2) transplants into large recipients weighing more than 100 kg, (3) grafts from females to males compared with males to females, (4) kidneys that experience rejection episodes, and (5) cadaveric grafts compared with living, unrelated donor grafts. A variety of factors have been suggested in the past to be responsible for the lower graft-survival rates of kidneys from infants [1, 16, 54, 97, 149], older donors [54, 107, 149]; or transplants into obese patients [16]. It has been clearly shown that, with age, there is a decrease in glomerular numbers and size [94] and, predictably, implants of such compromised kidneys reduced graft survival in the study by Terasaki's group [141]. Assuming that the extreme situation of a small kidney transplanted to a large recipient would occur more frequently in female to male than in male to female grafts, a difference in graft survival would be expected. This was confirmed by Yuge and Cecka [149].

It is reasonable to assume that rejection episodes result in permanent kidney damage with reduction in renal mass. Thus, rejection can be expected to increase the rate of hyperfiltration damage. While it has been clearly shown that episodes of acute rejection have a marked effect on subsequent graft survival rates, it is more difficult to distinguish the group in which hyperfiltration is contributing to long-term decline in graft function from the patients who are still immunologically active. Transplants from living, unrelated donors give greatly improved long-term graft-survival results compared with those from cadaveric donors. This has been suggested to be because of the greater 'nephron dose' in a healthy kidney than that in many cadaver kidneys [106, 129].

Further support for the nephron-dosing theory comes from Feehally's group [30], who noted that deteriorating kidney function, in five patients diagnosed as having chronic rejection, was arrested by simple dietary restrictions and no increase in immunosuppression. The similarity of pathological features of chronic allograft nephropathy and hyperfiltration has been reported [113] and the finding of a focal and segmental glomerulosclerosis is not unusual in a long-term graft [27].

7.3 Ischaemia–Reperfusion Damage

The detrimental role of oxygen free radicals in ischaemia/reperfusion-induced organ injury has been studied extensively in various organs [40, 75, 82, 100]. Experimental work has suggested an important role for free radicals generated from oxygen by activated xanthine oxidase at reperfusion in the cold-preserved heart and kidney transplant model [42, 50, 135]. Oxygen or hydroxyl free-radical damage to the vascular wall and the interstitial tissue may take place because of the continuing inflammatory process or through the formation of reactive species at the time of reperfusion. The ischaemia–reperfusion insult is known to cause endothelial damage, with a subsequent response to injury resulting in intimal hyperplasia. It has been shown both in experimental aortic grafts [146, 147] and in renal transplantation in man [65, 66] that the duration of ischaemia is important in predicting poor graft survival. Kidney allografts with prolonged warm ischaemia times show an increased frequency of acute rejection, and possibly, as described above, chronic vascular damage. Conversely, a decrease of ischaemia time significantly decreases later vascular and glomerular changes [85, 144].

Land's group (1994) showed that intraoperative treatment of kidney-transplant patients with human recombinant superoxide dismutase (SOD) for ablation of free radical-mediated reperfusion injury of cold-stored allogeneic cadaveric donor kidneys, significantly reduced the number of incidences of acute rejection as well as chronic rejection. Land proposed a working hypothesis for this beneficial effect based on the principle that the main target of free radical-mediated reperfusion injury is the microvascular endothelium. In his scheme, the injury has the potential to upregulate HLA-DR and adhesion molecule expression with an increase in phagocytic and inflammatory activity due to greater APC reactivity in the graft [38]. It can be seen that free radical-mediated or reperfusion-induced graft endothelial injury could be the key event contributing to the obliterative endarteritis seen in the longstanding, failing graft.

The beneficial effect of SOD on chronic rejection may be brought about by the primary prevention of acute rejection episodes, often associated histologically with endothelitis. Repeated acute rejection-mediated endothelial injury may, in this way, contribute to chronic obliterative endarteritis.

7.4 Hyperlipidaemia

Lipoprotein abnormalities may also influence the long-term fate of the graft. In experimental heart transplantation in the rat and rabbit, an increased dietary intake of cholesterol will accelerate the process of chronic rejection of the graft [33].

Hyperlipoproteinaemia has an established role in the development of naturally occurring atherosclerosis [110, 116, 132]. It has also been shown that there is a strong correlation between the degree of hyperlipidaemia and the extent of histopathological changes seen in chronic allograft nephropathy [23]. In chronic rejection patients, the same group have demonstrated higher levels of lipoproteins with atherogenic patterns than patients who have stable graft function. They also found that pre-transplant hypercholesterolaemia had a significant influence on graft function [24] and on graft losses due to chronic vascular rejection. The mechanism for this is thought to be oxidation of low-density lipoprotein (LDL) in a graft with an ongoing inflammatory process, oxidised LDL having the ability to induce class-II antigen expression [34]. Oxidised LDL may cause mononuclear inflammatory cells present within the graft to release a factor that stimulates expression of ICAM-1, vascular adhesion molecule 1 (VCAM-1) and endothelial cell adhesion molecule 1 (ECAM-1). This would increase adhesion of monocytes to endothelial cells [35], with increased class-I antigen expression on human monocytes [34], leading to increased expression of PDGF-AA transcripts in smooth muscle cells (SMCs) and expression of PDGF receptors on SMCs and a subsequent enhanced responsiveness of SMCs to PDGF [136]. Thus, the oxidation processes in the vessel wall and the oxidative modification of LDL may play an important role in the development of intimal hyperplasia, characteristic of chronic allograft nephropathy.

7.5 Hypertension

It is well established that hypertension damages endothelial cells and alters their function, as well as having effects on mitogenic expression in SMCs. Vasoactive peptides, such as angiotensin II, endothelin 1 (ET-1) and thromboxane A2, are also direct mitogens or act together with other growth factors on SMCs. Thromboxane enhances immune reactivity, stimulates platelet aggregation and causes vasoconstriction and SMC proliferation. Early results suggest that hypertension 1 year following grafting or pre-transplant hypertension affect subsequent loss of grafts due to chronic allograft nephropathy.

7.6 The Effect of Cytomegalovirus on the Renal Allograft

Cytomegalovirus (CMV) infection is an important cause of morbidity and mortality among allograft recipients, described by Dummer et al. [26] in heart or heart–lung transplants who received cyclosporin. In the cardiac-allograft patient, there is also strong clinical evidence implicating CMV in the arteriosclerosis characteristic of the long-term allograft [41, 71, 74]. CMV nucleic acids have been identified in the coronary arteries of heart-transplant recipients, with severe accelerated allograft arteriosclerosis [51]. The possible role of herpes viruses [29], in particular CMV [46, 47, 80], in the pathogenesis of native atherosclerosis has now been substantiated in recent studies [127].

It was Fabricant et al. [29] who demonstrated that, the herpes virus, that caused Marek's disease in chickens, induced athersclerotic lesions closely resembling those observed in classical atherosclerosis in human arteries. Lemstrom et al. [70], using a rat model and aortic allografts, demonstrated that CMV infection accelerated allograft arteriosclerosis. Infection at the time of transplantation was associated with a prominent early inflammatory episode and proliferation of inflammatory cells in the allograft adventitia, and was also shown to double the proliferation of SMCs within the intima. In renal allografts, the most striking feature is a CMV-associated glomerulopathy [49], similar in appearance to acute allograft glomerulitis, which may be modified or induced in some recipients by CMV infection [13, 78]. In latent CMV infection of SMCs [80, 81], the virus may transform the cells by incorporating into the cell genome, inducing local proliferation and production of growth factors [7, 43, 81].

Studies on acute CMV infection in mice suggest an important role for natural killer cells [119] and it was also noted that two kinds of cytotoxic T cells were produced against immediate and late antigens [108, 109]. In company with other herpes viruses, CMV induces the expression of IgG Fc receptors on the virus-infected cells [59], allowing interaction with granulocytes [76], leading to cell damage through enzyme and free-radical release. Human CMV infection upregulates IL-1β gene expression, further enhancing inflammatory responses [55], along with upregulation of PDGF-BB, TGF-Pl, ICAM-1 and lymphocyte function-associated antigen. The immediate early gene of the human virus codes for a protein with sequence homology and immunological cross reactivity with HLA-DR P chain, possibly enhancing alloimmune responses to donor antigens [36]. CMV also encodes a glycoprotein homologous to the heavy chain of MHC class-I antigens that has the ability to bind to P2 microglobulin [6]. Reinke et al. [111] showed that in a sub-population of renal transplant patients with a late rejection episode, antiviral drugs were effective in reversing the increased creatinine level. The peripheral blood of these individuals has been shown to contain large numbers of memory-type CD8+ cells.

Systemic infections of various types have been shown to be associated with the development of chronic allograft nephropathy [2]. The study found that CMV, other viruses and bacteria were all risk factors and suggested that, although infection and rejection are entirely different clinical events and necessitate different interventions, it is clear that the body's response to these two events is similar. Both allo- and microbial antigens stimulate the secretion of similar cytokines, namely TNF, IFN-γ and IL-6.

8 Pathophysiology of the Vascular Lesion in Chronic Allograft Nephropathy

The vascular changes in chronic allograft nephropathy are chiefly represented by narrowing of the arteriolar lumen secondary to a proliferative endarteritis that progressively obliterates and causes the vessels to undergo fibrosis. In another pattern, there is mononuclear infiltration of the intima associated with vascular smooth muscle proliferation. These cells migrate from the media to the intima, along with modifications of the arterial endothelium and intimal deposition of extracellular matrix. At an advanced stage, diffuse arteriolar obliteration is found along with massive interstitial fibrosis, tubular destruction and extensive glomerulosclerosis. The aetiology is multifactorial and complex and shares many features with other inflammatory responses involving the cytokine–adhesion molecule cascade.

However, there is growing evidence that the primary event in the development of this obliterative arteriosclerosis occurs as a 'response to injury', thought to be the underlying mechanism for the development of generalised atheroma. Primary endothelial injury would be the key event. This theory, formulated by Ross and Glomset in 1976 [117], and modified by Ross in 1986 [116], states that the lesions of atherosclerosis are initiated as a response to some form of injury to the arterial endothelium. In chronic allograft nephropathy, there may be repetitive endothelial injury or endothelial activation with intimal proliferation, hypertrophy and repair.

On a cellular level, the process in the vascular wall is characterised by migration of SMCs into the intima, with subsequent proliferation of these cells driven by paracrine or autocrine growth-factor stimulation [147]. The most important chemoattractants for SMCs are PDGF, TGF-, IL-, and insulin-like growth factor type I (IGF-1), which may be involved in the migration of SMCs from the media to the intima [32, 45, 103]. Proliferation of SMCs may be induced directly by PDGF, ETA and thrombin, or indirectly through the action of IL-, TNF- (X, TGF-P, vasoconstrictive substances (ET-, angiotensin II, thromboxane A2) or oxidised LDL.

The number of cytokines and growth factors possibly involved in tissue remodelling of chronic allograft nephropathy is large. There is certainly evidence that PDGF plays an important role in the development of the chronic vascular changes [44], as in naturally occurring atherosclerosis. It may be secreted by a number of cells, including monocytes/macrophages [120], endothelial cells [22], vascular SMCs and glomerular mesangial cells [124], or may be released when platelets aggregate or become activated in the vessel wall. PDGF has been demonstrated to be present in renal transplants undergoing chronic vascular rejection and in experimental heart and aorta transplants in the rat, simulating chronic vascular rejection. Furthermore, it has been demonstrated that PDGF-P receptors may be upregulated in tissues in which there is a chronic inflammatory process, such as chronic vascular rejection [31], as well as in experimental glomerulosclerosis. The expression of PDGF has been shown to be regulated by certain cytokines, vasoactive peptides (angiotensin II) and oxidised LDL [136].

Endothelial cell damage leads to reduced synthesis of endothelial-derived relaxing factor or nitric oxide (NO) [73], which was demonstrated following exposure to oxygen free radicals at reperfusion [105]. Due to the anti-proliferative

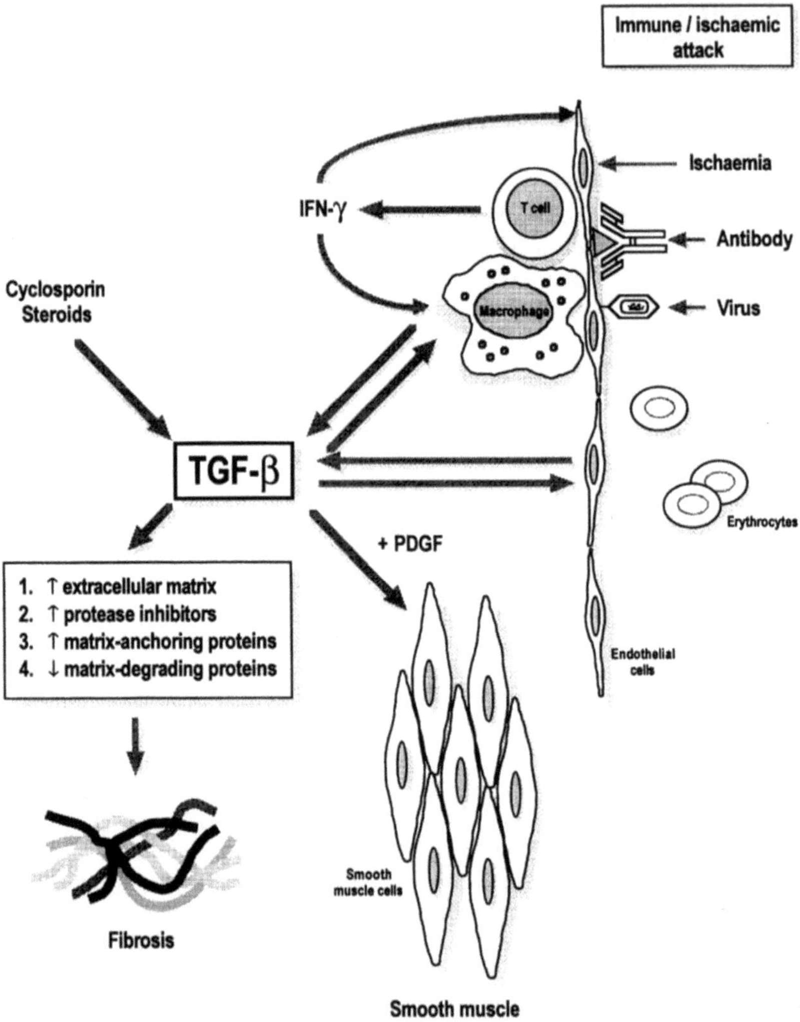

Fig. 2. Mechanism of chronic graft rejection. *T cell*, T lymphocyte; *IFN*, interferon; *TGF*, transforming growth factor; *PDGF*, platelet-derived growth factor

properties of NO, this may further enhance graft atherosclerosis. The damaged endothelial cell will release cytokines, including IFN-γ, IL- and TNF-cc, increase expression of adhesion molecules ICAM-I, VCAM-I and ELAM–1 [101, 128], upregulate class-II expression and become more permeable to plasma proteins. The cytokine gradient, thus established, favours leucocyte adhesion and activation. Finally, neutrophils, lymphocytes and macrophages migrate to the area and enter the surrounding tissues. The activated cells, particularly macrophages, elaborate cytokines and growth factors that increase smooth muscle proliferation, leading to eventual luminal obliteration.

Recent work has suggested that TGF- has a key role in the genesis of the two elements of chronic allograft nephropathy, namely SMC proliferation and fibrosis [118] (Fig. 2). TGFPI is a secretory product of many cell lines, including platelets, T cells, mesangial cells and monocytes/macrophages, and is a multifunctional cytokine [114]. It is a prominent member of the cytokine cascade involved in tissue repair, and considerable data support its fibrogenic properties [9]. Sharma's group [118] showed that intragraft expression of TGF–PI is a significant molecular correlate of interstitial fibrosis and chronic allograft nephropathy. They suggested that the specificity of this association was emphasised by the lack of correlation between histological features of either interstitial fibrosis or chronic allograft nephropathy and intragraft demonstration of type-I cytokines (IL-, IFN-γ), type-II cytokines (IL-, IL-), and cytotoxic attack molecules (granzyme B and perforin).

9 Summary and Future Prospects

Chronic allograft nephropathy is a major threat against long-term function and survival of transplanted kidneys. The pathogenesis is complex and multifactorial and involves immune processes against vascular and extravascular tissue components, perfusion failure and an array of haemodynamic and metabolic adaptations that occur in response to loss of organ mass. Today, there is no established treatment, although greater understanding of the pathogenesis has enabled new therapeutic strategies to be identified. The first strategy would be the optimisation of immunosuppressive treatment and development of new immunosuppressive agents. Next, metabolic intervention, in the form of lipid-lowering agents, antioxidants and thromboxane antagonists is of great importance. Third, prompt treatment or prevention of CMV and other infections is now recognised as vital for long-term graft survival. Finally, and perhaps the most exciting, is the development of new drugs to inhibit SMC proliferation. It is hoped that, although graft destruction appears inevitable, it may be possible in the future to reduce the intensity of its progression.

References

1. Alexander SR, Arbus GS, Butt KM, Conley S, Fine RN, Greiter I, Gruskin AB, Harmon WE, McEnery PT, Nevins TE, et al. (1990) The 1989 report of North American Pediatric Renal Transplant Cooperative Study. Pediatr Nephrol 4: 542–553
2. Almond PS, Matas AJ, Gillingham K, Dunn DL, Payne WD, Gores P, Gruessner R, Najarian JS (1993) Risk factors for chronic rejection in renal allograft recipients. Transplantation 55: 752–757
3. Batchelor JR (1980) The immunogenic signal of allografts. Adv Nephrol Necker Hosp 9: 237–244
4. Baltzan MA, Baltzan RB, Baltzan BL, Cunningham TC, Pylypchnk GB, Dyck RF, West ML (1990) HLA matching enhances long-term renal graft survival but does not relate to acute rejection. Medicine (Baltimore) 64: 227–231
5. Basadonna GP, Matas AJ, Gillingham KJ, Payne WD, Dunn DL, Sutherland DE, Gores PF, Gruessner RW, Arrazola L, Najarian JS (1993) Relationship between early vs late acute rejection and onset of chronic rejection in kidney transplantation. Transplant Proc 25: 910–911

6. Beck S, Barrell BG (1988) Human cytomegalovirus encodes a glycoprotein homologous to MHC class-antigens. Nature 331: 269–272
7. Benditt EP, Barrett T, Dougall JK (1983) Viruses in the etiology of atherosclerosis. Proc Natl Acad Sci U S A 80: 6386–6389
8. Besbas N, Sayed-Ahmed N, Cope GH (1992) Distribution of immunoreactive growth factors in ischemic renal injury. J Am Soc Nephrol 3: 703
9. Border WA, Nobel NA (1994) Transforming growth factor β in tissue fibrosis. N Engl J Med 331: 1286–1292
10. Braun MY, McCormack A, Webb G, Batchelor JR (1993) Mediation of acute but not chronic rejection of MHC-incompatible rat kidney grafts by alloreactive CD4 T cells activated by the direct pathway of sensitization. Transplantation 55: 177–182
11. Brenner BM, Meyer TW, Hostetter TH (1982) Dietary protein intake and the progressive nature of kidney disease: the role of hemodynamically mediated glomerular injury in the pathogenesis of progressive glomerular sclerosis in aging, renal ablation and intrinsic renal disease. N Engl J Med 11: 652–659
12. Brewer Y, Palmer A, Taube D, Welsh K, Bewick M, Bindon C, Hale G, Waldmann H, Dische F, Parsons V, et al. (1989) Effect of graft perfusion with two CD45 monoclonal antibodies on the incidence of kidney allograft rejection. Lancet ii: 935–937
13. Burdick JF, Beschorner WE, Smith WJ, McGraw DJ, Bender WL, Williams GM, Solez K (1984) Characteristics of early routine renal allograft biopsies. Transplantation 38: 1058–1067
14. Busch GJ, Schamberg JF, Morets RL, Strom TB, Tilney NL, Carpenter CB (1976) T and B cell patterns in irreversibly rejected human renal allografts: correlation of morphology with surface markers and cytotoxic capacity of the isolated lymphoid infiltrate. Lab Invest 35: 272–280
15. Cecka JM (1994) Outcome statistics of renal transplants with an emphasis on long-term survival. Clin Transplant 8: 324–327
16. Cecka JM, Terasaki PI (1990) Matching kidneys for size in renal transplantation. Clin Transplant 4: 82
17. Cook DJ, Terazki PI (1989) Renal transplantation in the American continent. In: Brent L, Sells RA (eds) Organ transplantation: current clinical and immunological concepts. Balliere, London, p 195
18. Cosio FG, Orosz CG (1992) Adhesion molecules and the kidney in health and disease. J Nephrol 6: 22–31
19. Cramer DV, Qian S, Harnaha J, Chapman FA, Starzl TE, Makowka L (1989) Accelerated graft arteriosclerosis is enhanced by sensitization of the recipients to donor lymphocytes. Transplant Proc 21: 3714–3715
20. Croker BP, Salomon DR (1989) Pathology of the renal allograft. In: Tisher CC, Brenner BM (eds) Renal pathology with clinical and functional correlations. Lippincott, Philadelphia, pp 1518–1554
21. D'Ardenne AJ, Dunnill MS, Thompson JF, McWhinnie D, Wood RFM, Morris PJ (1986) Cyclosporin and renal graft histology. J Clin Pathol 39: 145–151
22. DiCorleto PE, Bowen-Pope DF (1983) Cultured endothelial cells produce a platelet-derived growth factor-like protein. Proc Natl Acad Sci U S A 80: 1919–1923
23. Dimeny E, Fellstrom B, Larsson E, Tufveson G, Lithell H (1993a) Hyperlipoproteinaemia in renal transplant recipients – is there a linkage with chronic vascular rejection? Transplant Proc 25: 2065–2066
24. Dimeny E, Tufveson G, Lithell H, Larsson E, Siegbahn A, Fellstrom B (1993b) The influence of pretransplant lipoprotein abnormalities on early results of renal transplant. Eur J Clin Invest 23: 572–579
25. Doxiadis IIN, Smits JMA, Schreuder GMTh, Persijn GG, van Houwelingen HC, van Rood JJ, Claas FHJ (1996) Association between specific HLA combinations and probability of kidney allograft loss: the taboo concept. Lancet 348: 850–853
26. Dummer JS, White LT, Ho M, Griffith BP, Hardesty RL (1985) Morbidity of cytomegalovirus infection in recipients of heart or heart-lung transplants who received cyclosporine. J Infect Dis 152: 1182–1191
27. Dunnill MS (1984) Histopathology of rejection in renal transplantation. In: Dunnill MS (ed) Kidney transplantation. Gruen and Stratton, New York, p 355

28. Dustin ML, Springer TA (1991) Role of lymphocyte adhesion receptors in transient interactions and cell locomotion. Ann Rev Immunol 9: 27–66
29. Fabricant CG, Fabricant J, Litrenta MM, Mimick CR (1978) Virus-induced atherosclerosis. J Exp Med 148: 335–340
30. Feehally J, Harris KP, Bennett SE, Walls J (1986) Is chronic renal transplant rejection a non-immunological phenomenom? Lancet ii: 486–488
31. Fellstrom B, Klareskog L, Terracio L (1989) Platelet derived growth factor receptors in the kidney – importance of up-regulated expression in renal inflammation. Kidney Int 36: 1099–1102
32. Fellstrom B, Larsson E (1993) Pathogenesis and treatment perspectives in chronic graft rejection. Immunol Rev 134: 83–98
33. Fellstrom B, Dimeny E, Larsson E, Claesson K, Tufveson G (1990) Rapidly proliferative arteriopathy in cyclosporin-induced permanently surviving rat cardiac allografts simulating chronic vascular rejection. Clin Exp Immunol 80: 288–292
34. Frostegard J, Nilsson J, Haegerstrand A, Hausten A, Wiazell H, Gidlund M (1990) Oxidised low density lipoprotein induces differentiation and adhesion of human monocytes and the monocytic cell line U937. Proc Natl Acad Sci U S A 87: 904–908
35. Frostegard J, Haegerstrand A, Gidlund M, Nilsson J (1991) Biologically modified low density lipoprotein increases the adhesive properties of vascular endothelial cells. Atherosclerosis 90: 119–126
36. Fujinami RS, Nelson JA, Walker L, Oldstone MB (1988) Sequence homology and immunologic cross reactivity of human cytomegalovirus with HLA-DR β chain: a means for graft rejection and immunosuppression. J Virol 62: 100–105
37. Garovoy MR, Reddish MA, Busch GJ, Tilney NL (1982) lmmunoglobulin secreting cells recovered from rejected human renal allografts. Transplantation 33: 109–111
38. Gasic AC, McGuire G, Krater S, Farhood Al, Goldstein MA, Smith CW, Entman ML, Taylor AA (1991) Hydrogen peroxide pretreatment of perfused canine vessels induces ICAM-1 and CD18-dependent neutrophil adherence. Circulation 84: 2154–2166
39. Goldberg LC, Cook T, Taube D (1994) Pretreatment of renal allografts with anti-CD45 antibodies: optimisation of perfusion technique. Transplant Immunol 2: 27–34
40. Granger DN, Hollwarth ME, Parks DA (1986) Ischemia-reperfusion injury: role of oxygen derived free radicals. Acta Physiol Scand 548: 47–63
41. Grattan MT, Moreno-Cabral CE, Starnes VA, Oyer PE, Stinson EB, Shurnway NE (1989) Cytomegalovirus infection is associated with cardiac allograft rejection and atherosclerosis. JAMA 261: 3561–3566
42. Green CJ, Healing G, Lunec J, Fuller BJ, Sirnkin S (1986) Evidence of free radical-induced damage in rabbit kidneys after single hypothermic preservation and autotransplantation. Transplantation 41: 161–165
43. Hajjar DP (1991) Viral pathogenesis of atherosclerosis: impact of molecular mimicry and viral genes. Am J Pathol 139: 1195–1211
44. Hayry P, Yilmaz S (1995) The role of growth factors in graft vessel disease. Transplant Proc 27: 2066–2067
45. Hayry P, Isoniemi H, Yilmaz S, Mennander A, Lemstrom K, RaisanenSokalowski A, Kosinen P, Ustinov J, Lautenschlager 1, Taskinen E, et al. (1993) Chronic allograft rejection. Immunol Rev 134: 33–81
46. Hendrix MGR, Dormans PHJ, Kitslaar P, Bosman F, Bruggeman CA (1989) The presence of cytomegalovirus nucleic acids in arterial walls of atherosclerotic and non-atherosclerotic patients. Am J Pathol 134: 1151–1157
47. Hendrix MGR, Salimans MMM, van Boven CPA, Bruggeman CA (1990) High prevalence of latently present cytomegalovirus in arterial walls of patients suffering from grade III atherosclerosis. Am J Pathol 136: 23–28
48. Hill GS, Rosengard B, Benedict Cl, Sachas DH (1990) Progressive glomerular hypertrophy and hyperfiltration glomerulosclerosis in chronic renal allograft rejection (abstract). J Am Soc Nephrol 1: 749
49. Hiki Y, Leong ASY, Mathew TH, Seymour AE, Pascoe V, Woodroffe AJ (1986) Typing of intraglomerular mononuclear cells are associated with transplant glomerular rejection. Clin Nephrol 26: 244–249

50. Hoshino T, Maley WR, Bulkley GB, Williams GM (1988) Ablation of free radical-mediated reperfusion injury for the salvage of kidneys taken from non-heart beating donors. Transplantation 45: 284–289
51. Hruban RH, Wu T-C, Beschorner WE, Cameron DE, Ambinder RF, Baumgartner WA, Hutchins GM, Reitz BA (1990) Cytomegalovirus nucleic acids in allografted hearts. Hum Pathol 21: 981–983
52. Hutchinson IV (1988) Specific immunosuppression. In: Morris PJ (ed) Kidney transplantation. Principles and practice. Saunders, Philadelphia, pp 389–416
53. Hutchinson IV (1993) Effector mechanisms in transplant rejection – an overview. In: Rose ML, Yacoub MH (eds) Immunology of heart and lung transplantation. Edward Arnold, Boston, pp 3–21
54. Ito Y, Iwaki Y, Terasaki PI (1986) Donor and recipient age effect. In: Terasaki PI (ed) Clinical transplants. UCLA Tissue Typing Laboratory, Los Angeles, p 189
55. Iwamoto GK, Monick MM, Clark BD, Auron PE, Stinski MF, Hunninghake GW (1990) Modulation of interleukin IP gene expression by the immediate early genes of human cytomegalovirus. J Clin Invest 85: 1853–1857
56. Joseph JV, Guy SP, Brenchley PEC, Parrott NR, Short CD, Johnson RW, Hutchinson IV (1995) Thl and Th2 cytokine gene expression in human renal allografts. Transplant Proc 27: 915–916
57. Kasiske BL, Kalil RSN, Lee HS, Raoke V (1991) Histopathological findings associated with a chronic progressive decline in renal allograft function. Kidney Int 40: 514–524
58. Keane WF, Kasiske BL, O'Donnell MP (1988) Lipids and progressive glomerulosclerosis. Am J Nephrol 8: 261–271
59. Keller RR, Peitchel JN, Goldman JN, Goldman M (1976) An IgG–Fc receptor induced in cytomegalovirus-infected human fibroblasts. J Immunol 116: 772–777
60. Kemeny DM, Noble A, Holmes BJ, Diaz-Sanchez D (1994) Immune regulation: a new role for the CD8+ T cell. Immunol Today 15: 107–110
61. Keogh A, Kaan A, Doran T, Macdonald P, Bryant D, Spratt P (1995) HLA mismatching and outcome in heart, heart–lung and single lung transplantation. J Heart Lung Transplant 14: 444–451
62. Kingma I, Chea R, Davidoff A, Grothman G, Benediktsson H, Paul LC (1990) Glomerular haemodynamics and focal glomerulosclerosis in chronic renal allograft rejection in the rat (abstract). J Am Soc Nephrol 1: 749
63. Kirkman RL, Strom TB, Weir MR, Tilney NL (1982) Late mortality and morbidity in recipients of long-term renal allografts. Transplantation 34: 347–351
64. Knight RJ, Kerman RH, Welsh M, Golden D, Schoenberg L, Van Buren CT, Lewis RM, Kahan BD (1991) Chronic rejection in primary renal allograft recipients under cyclosporine–prednisone immunosuppressive therapy. Transplantation 51: 355–359
65. Deleted in production
66. Land W, Schneeberger H, Schleibner S, Illner WD, Abendroth D, Rutili G, Arfors KE, Messmer K (1994b) Beneficial effect of human recombinant superoxide dismutase on acute and chronic rejection events in recipients of cadaveric renal transplants. Transplantation 57: 211–217
67. Lau M, Terasaki PI, Park MS, Barbetti A (1991) International cell exchange. In: Terasaki PI (ed) Clinical transplants. UCLA Tissue Typing Laboratory, Los Angeles, pp 385–400
68. Lechler RI, Batchelor JR (1982) Restoration of immunogenicity to passenger cell-depleted kidney allografts by the addition of donor strain dendritic cells. J Exp Med 155: 31–41
69. Legendre C, Droz D, Saltiel C (1990) Correlations between chronic rejection, pathologic lesions and HLA-class I and II compatibilities. In: Touraine JL, Traeger J, Betuel H (eds) Transplantation and clinical immunology XXI. Elsevier, New York, pp 157–165
70. Lemstrom KB, Bruning JH, Bruggeman CA, Lautenshlager IT, Hayry PJ (1993) Cytomegalovirus infection enhances smooth muscle cell proliferation and intimal thickening of rat aortic allografts. J Clin Invest 92: 549–558
71. Loebe M, Schuler S, Ortwin Z, Warnecke E, Fleck E, Hetzer R (1990) Role of cytomegalovirus infection in the development of coronary artery disease in the transplanted heart. J Heart Transplant 9: 707–711

72. Lowry RP (1993) The relationship of IL-4, IL-10 and other cytokines to transplant tolerance. Transplant Sci 3: 104–112
73. Luscher TF (1990) The endothelium. Target and promoter of hypertension? Hypertension 15: 482–485
74. McDonald K, Rector CS, Braunlin EA, Kubo SH, Olivari MT (1989) Association of coronary artery disease in cardiac transplant recipients with cytomegalovirus infection. Am J Cardiol 64: 359–362
75. McCord JM (1985) Oxygen-derived free radicals in post ischemic tissue injury. N Engl J Med 312: 159–163
76. MacGregor RR, Friedman HM, Macarak EJ, Kefalides NA (1980) Virus infection of endothelial cells increases granulocyte adherence. J Clin Invest 65: 1469–1477
77. Mason DW, Morris PJ (1986) Effector mechanisms in allograft rejection. Annu Rev Immunol 4: 119–145
78. Matas AJ, Sibley R, Mauer SM, Kim Y, Sutherland DER, Simmons RL, Najarian JS (1982) Pre-discharge, post-transplant kidney biopsy does not predict rejection. J Surg Res 32: 269–274
79. Maruya E, Takemoto S, Terasaki PI (1994) HLA matching: identification of permissible HLA mismatches. In: Terasaki PI, Cecka JM (eds) Clinical transplants. UCLA Tissue Typing Laboratory, Los Angeles, pp 285–292
80. Melnick JL, Dreesman GR, McCollum CH, Petrie BL, Burek J, DeBakey ME (1983) Cytomegalovirus antigen within human arterial smooth muscle cells. Lancet ii: 644–647
81. Melnick JL, Adam E, DeBakey ME (1990) Possible role of cytomegalovirus in atherogenesis. JAMA 262: 2204–2207
82. Menger MD, Lehr HA, Messmer K (1991) Role of oxygen radicals in the microcirculatory manifestations of postischemic injury. Klin Wochenschr 69: 1050–1055
83. Modena FM, Hostetter TH, Salahudeen AK, Najarian JS, Matas AJ, Rosenberg ME (1991) Progression of kidney disease in chronic renal transplant rejection. Transplantation 52: 239–244
84. Moretta L, Ciccone E, Moretta A, Hoglund P, Oulen C, Karre K (1992) Allorecognition by NK cells: nonself or no self? Immunol Today 13: 300–306
85. Munger KA, Coffman TM, Griffiths RC, Fogo A, Badr KF (1983) Determinants of long-term renal allograft survival. Transplantation 55: 1219–1224
86. Naimark DMJ, Cole E (1994) Determinants of long-term renal allograft survival. Transplant Rev 8: 93–113
87. Najarian JS, Fryd DS, Strand M, Canafax DM, Ascher NL, Payne WD, Simmons RL, Sutherland DE (1985) A single institution, randomized, prospective trial of cyclosporine versus azathioprine–antilymphocyte globulin for immunosuppression in renal allograft recipients. Ann Surg 201: 142–157
88. Nasasdy T, Ormos J, Stiller D, Csajbak E, Szenohradsky P (1988) Tubular ultrastructure in rejected human renal allografts. Ultrastruct Pathol 12: 195–207
89. Nath KA (1993) The role of tubulo-interstitial processes in progressive renal disease. In: El Naha AM, Mallick NO, Anderson S (eds) Prevention of progressive renal failure. Oxford University Press, Oxford, pp 62–97
90. Nath KA, Hostetter MK, Hostetter TH (1985) Pathophysiology of chronic tubulo-interstitial disease in rats: interactions of dietary acid load, ammonia and complement component C3. J Clin Invest 76: 667–675
91. Nathan CF, Murray HW, Cohn ZA (1980) The macrophage as an effector cell. N Engl J Med 303: 622–626
92. Neild GH, Taube DH, Hartley RB, Bignardi L, Cameron JS, Williams DG, Ogg CS, Rudge CJ (1986) Morphological differentiation between rejection and cyclosporine nephrotoxicity in renal allografts. J Clin Pathol 39: 152–159
93. Novick AC, Gephardt G, Guz B, Steinmuller D, Tabbs RR (1991) Longterm follow-up after partial removal of a solitary kidney. N Engl J Med 325: 1058–1062
94. Nyengaard JR, Bendtsen TF (1992) Glomerular number and size in relation to age, kidney weight and body surface in normal man. Anat Rec 232: 194–201

95. Oguma S, Belle S, Starzl TE, Demetris AJ (1989) A histometric analysis of chronically rejected human liver allografts: insights into the mechanisms of bile duct loss: direct immunologic and ischemic factors. Hepatology 9: 204–209
96. Opelz G (1984) Blood transfusions and renal transplantation. In: Morris PJ (ed) Kidney transplantation. Principles and practice. Grune and Stratton, Orlando, Florida, pp 323–324
97. Opelz G (1988) Influence of recipient and donor age on pediatric renal transplantation. Transplant Int 1: 95–98
98. Opelz G, for the Collaborative Transplant Study (1991) Strength of HLAA, HLA-B and HLA-DR mismatches in relation to short-and long-term kidney graft survival. Transplant Int 5[Suppl]: 621
99. Orosz CG (1994) Endothelial activation and chronic allograft rejection. Clin Transplant 8: 299–303
100. Parks DA, Granger DN (1988) Ischemia-reperfusion injury: a radical view. Hepatology 8: 680–682
101. Patarroyo MP, Prieto J, Rincon J, Timonen T, Lundberg C, Lindborn L, Asjo B, Gahmberg CG (1990) Leukocyte-cell adhesion: a molecular process fundamental in leukocyte physiology. Immunol Rev 114: 67–108
102. Paul LC, Benediktsson H (1995) Post-transplant hypertension and chronic renal allograft failure. Kidney Int 52[Suppl]: 34–37
103. Paul LC, Fellstrom B (1992) Chronic vascular *rejection of the heart and kidney. Have rational treatment options emerged? Transplantation 53: 1169–1179
104. Paul LC, Hayry P, Foegh M, Dennis MJ, Mihatsch MJ, Larsson E, Fellstrom B (1993) Diagnostic criteria for chronic rejection/accelerated graft atherosclerosis in heart and kidney transplants: joint proceedings from the Fourth Alexis Carrel Conference on Chronic Rejection and Accelerated Arteriosclerosis in Transplanted Organs. Transplant Proc 25: 2022–2023
105. Pinsky DJ, Oz MC, Koga S, Taha Z, Broekman MJ, Marcus AJ, Liao H, Naka Y, Brett J, Cannon PJ, et al. (1994) Cardiac preservation is enhanced in a heterotopic rat transplant model by supplementing the nitric oxide pathway. J Clin Invest 93: 2291–2297
106. Pirsch JD, Sollinger HW, Kalayoglu M, Stratta RJ, D'Alessandro AM, Armbrush MJ, Belzer FO (1988) Living unrelated renal transplantation: results in 40 patients. Am J Kidney Dis 12: 499–503
107. Rao KV, Kasiske BL, Odlund MD, Ney AL, Anderson RC (1990) Influence of cadaver donor age on posttransplant renal function and graft outcome. Transplantation 49: 91–95
108. Reddehase MJ, Keil GM, Koszinowski UH (1984a) The cytolytic T lymphocyte response to the murine cytomegalovirus. 1. Distinct maturation stages of cytolytic T lymphocytes constitute the cellular immune response during acute infection of mice with the murine cytomegalovirus. J Immunol 132: 482–489
109. Reddehase MJ, Keil GM, Koszinowski UH (1984b) The cytolytic T lymphocyte response to the murine cytomegalovirus. 11. Detection of virus replication stage-specific antigen by separate populations of in vivo active cytolytic T lymphocyte precursors. Eur J Immunol 14: 56–61
110. Regnstrom J, Nilsson J, Tornvall P, Landou C, Hamsten A (1992) Susceptibility of low-density lipoprotein oxidation and coronary atherosclerosis in man. Lancet 339: 1183–1186
111. Reinke P, Fietze E, Ode-Hakim S, Prosch S, Lippert J, Ewert R, Volk HD (1994) Late acute renal allograft rejection and symptomless cytomegalovirus infection. Lancet 344: 1737–1738
112. Remuzzi G, Bertani T (1990) Is glomerulosclerosis a consequence of altered glomerular permeability to macromolecules? Kidney Int 38: 384–394
113. Rennke HG (1986) Pathology of glomerular hyperfiltration. In: Mitch WE, Brenner BM, Stein JH (eds) The progressive nature of renal disease. Churchill-Livingstone, New York, p 111 (Contemporary Issues in Nephrology, vol 14)
114. Roberts AB, Sporn MB (1993) Physiological actions and clinical applications of transforming growth factor-β (TGF-β). Growth Factors 8: 1–9

115. Rose A, Uys CJ (1990) Pathology of graft atherosclerosis (chronic rejection). In: Cooper DKC, Novitsky D (eds) Transplantation and replacement of thoracic organs. Kluwer, Boston
116. Ross R (1986) The pathogenesis of atherosclerosis – an update. N Engl J Med 314: 488–500
117. Ross R, Glomset JA (1976) The pathogenesis of atherosclerosis. N Engl J Med 295: 369–377
118. Sharma VK, Bologa RM, Xu GP, Li B, Mouradian J, Wang J, Serur D, Rao V, Suthanthiran (1996) Intragraft TGF-beta1 mRNA: a correlate of interstitial fibrosis and chronic allograft nephropathy. Kidney Int 49: 1297–1303
119. Shellarn GR, Allen JE, Papadimitriou JM, Bancroft GJ (1981) Increased susceptibility to cytomegalovirus infection in beige mutant mice. Proc Natl Acad Sci USA 78: 5104–5108
120. Shimokado K, Raines EW, Madtes DK, Barrett TB, Benditt EP, Ross R (1985) A significant part of macrophage-derived growth factor consists of at least two forms of PDGF. Cell 43: 277–286
121. Sibley RK (1994) Morphological feature of chronic rejection in kidney and less commonly transplanted organs. Clin Transplant 8: 293–298
122. Sibley RK, Rynasiewicz J, Ferguson RM, Fryd D, Sutherland DER, Simmons RL, Najarian JS (1983) Morphology of cyclosporine nephrotoxicity and acute rejection in patients immunosuppressed with cyclosporine and prednisolone. Surgery 94: 225–234
123. Sigal NH, Dumont FJ (1992) Cyclosporin A, FK506 and rapamycin: pharmacologic probes of lymphocyte signal transduction. Annu Rev Immunol 10: 519–560
124. Silver BJ, Jaffer FE, Abboud HE (1989) Platelet-derived growth factor synthesis in mesangial cells: induction by multiple peptide mitogens. Proc Natl Acad Sci U S A 86: 1056–1060
125. Solez K, Axelsen RA, Benediktsson H, Burdick JF, Cohen AH, Colvin RB, Croker BP, Droz D, Dunnill MS, Halloran PF, et al. (1993a) International standardisation of criteria for the histologic diagnosis of renal allograft rejection: the Banff working classification of kidney transplant pathology. Kidney Int 44: 411–422
126. Solez K, Racusen LC, Marcussen N, Slatnik 1, Keown P, Burdick JF, Olsen S (1993b) Morphology of ischemic acute renal failure, function, and cycosporine toxicity in cyclosporine-treated renal allogaft recipients. Kidney Int 43: 1058–1067
127. Span AHM, Grauls G, Bosman F, van Boven CPA, Bruggeman CA (1992) Cytomegalovirus infection induces vascular injury in the rat. Atherosclerosis 93: 41–52
128. Springer TA (1990) Adhesion receptors of the immune system. Nature 346: 425–434
129. Squifflet JP, Pirson Y, Poncelet A, Gianello P, Alexandre GP (1990) Unrelated living donor kidney transplantation. Transplant Int 3: 32–35
130. Starzi TE, Demetris AJ, Murase N, Ildstad S, Ricordi C, Trucco M (1992) Cell migration, chimerism, and graft acceptance. Lancet 339: 1579
131. Starzi TE, Demetris AJ, Trucco M, Zeevi A, Ramos H, Terasaki P, Rudert WA, Kocova M, Ricordi C, Ildstad S, Murase N (1993) Chimerism and donor-specific non-reactivity 27 to 29 years after kidney allotransplantation. Transplantation 55: 1272–1277
132. Steinberg D, Parthasarathy S, Carew TE, Khoo JC, Witzturn JL (1989) Beyond cholesterol. Modifications of low-density lipoprotein that increase its atherogenicity. N Engl J Med 320: 915–924
133. Steinman RM (1991) The dendritic cell system and its role in immunogenicity. Annu Rev Immunol 9: 271–296
134. Steinman RM, Cohn ZA (1973) Identification of a novel cell type in peripheral lymphoid organs of mice. 1. Morphology, quantitation, tissue distribution. J Exp Med 137: 1142–1162
135. Stewart JR, Frist WH, Merrill WH (1990) Oxygen scavengers in myocardial preservation during transplantation. Methods Enzymol 186: 742–748
136. Stiko-Rahm A, Hultgardh-Nilsson A, Regnstrom J, Hamsten A, Nilsson J (1992) Native and oxidised LDL enhances production of PDGF-AA and the surface expression of PDGF receptors in cultured human arterial smooth muscle cells. Arterioscler Thromb 12: 1099–1109
137. Strom TB, Tilney NL, Carpenter CB, Busch GJ (1975) Identity and cytotoxic capacity of cells infiltrating renal allografts. N Engl J Med 292: 1257–1263

138. Takemoto S, Terazaki PI, Cecka JM, Cho YW, Gjertson DW (1992) Survival of nationally shared, HLA-matched kidney transplants from cadaveric donors. N Engl J Med 327: 834–839
139. Terasaki PI, Cecka JM, Cho Y (1990) Overview. In: Terasaki PI (ed) Clinical transplants. UCLA Tissue Typing Laboratory, Los Angeles, pp 585–601
140. Terasaki PI, Cecka M, Gjertson DW, Takernoto S (1995) High survival rates of kidney transplants from spousal and living unrelated donors. N Engl J Med 333: 333–336
141. Terasaki PI, Koyama H, Cecka JM, Gjertson DW (1994) The hyperfiltration hypothesis in renal transplantation. Transplantation 57(10): 1450–1454
142. Thorogood J, Van Houwelingen HQ, Van Rood JJ, Zantroort FA, Schrender GM, Persijn GG (1992) Factors contributing to long term kidney graft survival in Eurotransplant. Transplantation 54: 152–158
143. Tullius SG, Tilney NL (1995) Both allo antigen-dependent and independent factors influence chronic allograft rejection. Transplantation 59: 313–318
144. Van Es A, Hermans J, Van Bockel JH, Persijn GG, van Hooff JP de Graeff J (1983) Effect of warm ischaemia time and HLA (Aand B) matching on renal cadaveric graft survival and rejection episodes. Transplantation 36: 255–258
145. van Rood JJ (1967) A proposal for international cooperation in organtransplantation. In: Curtoni ES, Mattiuz PL, Tosi RM (eds) Histocompatibility testing. Munksgaard, Copenhagen, pp 451–452
146. Wanders A, Akyurek NIL, Waltenberger J, Stafberg C, Larsson E, Zhiping R, Funa K, Fellstrom B (1993) The impact of ischemia time on chronic vascular rejection in the rat – effects of angiopeptin. Transplant Proc 25: 2098–2099
147. Wanders A, Akyurek ML, Waltenberger J, Ren ZP, Stafberg C, Funa K, Larsson E, Fellstrom B (1995) Ischemia induced transplant arteriosclerosis in the rat. Arterioscler Thromb 15: 145–155
148. Yoshida Y, Fogo A, Ichikawa I (1989) Glomerular haemodynamic changes vs hypertrophy in experimental glomerular sclerosis. Kidney Int 35: 654–660
149. Yuge J, Cecka JM (1992) Sex and age effects in renal transplantation. In: Terasaki PI (ed) Clinical transplants. UCLA Tissue Typing Laboratory, Los Angeles, pp 269

Pathology of the Liver Allograft

B. PORTMANN and G. KOUKOULIS

Current Topics in Pathology
Volume 92, C.L. Berry (Ed.)

1 Introduction

Liver transplantation is now a well established and widely used treatment for otherwise incurable liver diseases. Steady improvements in donor organ preservation, surgical techniques and immunosuppressive regimens have reduced the rate of complications and allowed many patients to live long after liver transplantation. Likewise, a wide use of diagnostic liver biopsies has increased the confidence of pathologists in interpreting allograft changes at all stages after surgery. In many instances, histological assessment remains the "gold standard" for diagnosis. Due to an increasing number of active liver-transplantation programs and prolonged patient survival, pathologists working in specialised centres as well as those working in referring hospitals may be confronted with liver-allograft biopsies. Therefore, basic knowledge of this topic is now of clinical and academic importance. In the following account, the various pathological changes that may affect liver allografts are reviewed with an emphasis on diagnostic difficulties.

2 Indication and Timing of Liver Biopsy

The indication and, in particular, the timing of a biopsy vary greatly throughout different centres, depending on the number of patients in therapy trials assessing new immunosuppressive or antiviral drugs. As a baseline, most groups will take a needle-biopsy specimen during surgery at some point after reperfusion of the graft in the recipient (the "time-zero" biopsy) and, if there are no contraindications, on day 7, irrespective of graft function or clinical condition of the patient (protocol biopsy) a second specimen is taken. Whenever possible, a protocol biopsy specimen is obtained at the 1-year follow-up and again at 5 years in clinically stable patients. This is done prior to considering withdrawal of immunosuppression. Outside of these scheduled biopsies, specimens are taken on clinical indication, including: (1) investigation of acute graft dysfunction, especially prior to modifying immunosuppressive therapy; (2) assessment of the response to a specific treatment and for the purpose of documenting intractable rejection prior to the use of a rescue therapy, such as FK506 (prograf); (3) investigation of a lesion of undetermined nature; and (4) assessment of the progression of an allograft pathology of known aetiology, such as ductopenic rejection, biliary stricture or chronic hepatitis (CH) [either idiopathic or due to recurrent hepatitis B virus (HBV) or hepatitis C virus (HCV)] [94]. Fine needle aspiration [60] may offer advantages, but the technique is demanding and only used routinely by a few centres.

3 Donor Liver: "Time-Zero" Biopsy

3.1 Pre-existing Donor Injury

Severe macro-vesicular steatosis was recognised early as a cause of primary non-function [97]. Large fat vacuoles are thought to interfere with the perfusion procedure, and intracellular lipid may activate phospholipases with free-radical formation, thus exacerbating any reperfusion injury [122]. Liver with fatty change affecting 60%–70% of hepatocytes should not be used [12]. Lesser degrees of steatosis may not cause significant graft dysfunction if not combined with other risk factors, but correlate with early elevation of transaminases and, in subsequent histology, with the formation of large sinusoidal spaces – lipo-peliosis [34].

A few iron-loaded livers from undiagnosed haemochromatotic donors have been inadvertently transplanted. This is usually inconsequential, with progressive diminution of the siderosis following transplantation [98]. Severe siderosis persisted in the rare instance in which the recipient had an underlying genetic defect in iron metabolism [58].

3.2 Preservation/Reperfusion Damage

Cold ischaemia has been shown to damage the sinusoidal endothelium with subsequent activation and platelet adhesion after reperfusion in the recipient [11]. In both experimental animals and humans, the changes due to cold ischaemia can only be detected by electron microscopy, and their impact on graft function and reversibility are uncertain. They are, therefore, of little clinical value in assessing graft viability as results would only be available after the graft had been inserted. Hepatocytes seem more sensitive to warm ischaemia, either related to pre-terminal events in the donor or during warming up and reperfusion in the recipient. Various abnormalities have been described in "time-zero" specimens, including hepatocyte ballooning, spotty necrosis, apoptotic bodies, microvesicular steatosis and neutrophilic polymorph aggregates. These are changes that may be seen in unused portions of donor livers. The changes identified in donor livers prior to transplantation are usually not predictive of graft failure, with the exception of macrovesicular steatosis, as already discussed. This is not surprising, considering that following cell death – which has no expression in conventional histology – several hours are needed in a normally perfused organ for cell necrosis to become histologically evident. As a consequence, pathologists should be aware that perivenular hepatocyte necrosis or drop-out observed in a day-7 biopsy specimen may result from reperfusion damage, although it was not yet evident in the "time-zero" specimen. Another common finding in reperfusion specimens is a heavy centrolobular lipofuscin accumulation, most likely the result of abnormal lipid oxidation.

4 Early Graft Changes Other than Rejection

4.1 Primary Graft Non-function

Primary non-function is defined as a situation where the graft completely fails to function in the presence of an apparently normal vascular supply and in the absence of an identifiable cause other than a likely preservation injury. It is characterised clinically by lack of bile production, a marked elevation of transaminase levels, severe coagulopathy, renal failure and acidosis in the immediate post-operative period. The incidence varies from very low to 10% of grafts in individual centres and urgent re-transplantation is required. When unrelated to severe fatty change, a "time-zero" biopsy in these patients may show diffuse fine vesiculation of hepatocytes associated with widespread neutrophilic infiltration and focal necrosis [54]. Removed grafts show either massive fatty infiltration with fatty-cyst formation or varying degrees of widespread spotty or confluent, often haemorrhagic, necrosis.

4.2 Acute Graft Failure/Massive Haemorrhagic Necrosis

4.2.1 Hepatic Artery Thrombosis

Hepatic artery thrombosis in the grafted liver exhibits a range of presentation from the most common rapid deterioration in liver function and appearance of encephalopathy within 10 days of the transplant [59] to a remarkable preservation of function, particularly in children in whom this complication has a higher incidence (9%–18% vs 5% in adult). This is often inconsequential in a non-transplant setting. Failed grafts show map-like areas of coagulative necrosis with haemorrhagic borders. In cases with a sub-clinical course, the findings of a positive blood culture for enteric organisms may first indicate this complication. It is uncommon for this to develop into major biliary complications or multiple liver abscesses secondary to ischemic cholangitis [42].

4.2.2 Apparent Patent Artery

In a small proportion of recipients, initial graft function is followed by sudden deterioration, progressing to acute liver failure within 3–15 days of transplantation. In the early series, kinking of the hepatic artery was thought to be responsible and the term "septic hepatic gangrene" was applied because of the common contamination of the dead tissue by gram-negative organisms. The non-committal terminology of massive haemorrhagic necrosis was used to acknowledge the haemorrhagic appearance of the graft in this situation [50]. There is now good evidence that a number of these cases are examples of antibody-mediated rejection or hyperacute rejection [115]. Occasionally, severe graft necrosis may represent hypovolaemic injury, to which the denervated liver seems to be more susceptible [49]. In addition, cases of haemorrhagic necrosis leading to early graft failure have been observed in patients transplanted for acute liver failure due to non-A non-B hepatitis. Electron microscopy of native liver and necrotic graft revealed toga-like viral particles, suggesting that a few cases of haemorrhagic necrosis might result from graft viral re-infection, possibly a single organ Shwarzman reaction [31].

Recently, high levels of interferon-γ and tumour necrosis factor-α were detected in five cases of haemorrhagic necrosis, suggesting that cytokine-mediated inflammatory response may lead to a univisceral Shwartzman reaction [6]. Diffuse hepatocytic changes with nuclear pyknosis or loss, and scattered eosinophilic necrotic cells without a zonal distribution may be seen on liver biopsy prior to the development of massive haemorrhagic necrosis. It must be noted that a well-demarcated infarct in a biopsy specimen is not always a forerunner of widespread graft ischaemia. Localised subcapsular necrosis or, in reduced grafts, necrosis along the line of surgical cut-down may have been sampled and are usually of no consequence.

4.3 "Functional" Cholestasis and Hepatocyte Ballooning

Perivenular cholestasis is a common finding after transplantation and, in itself, is of little diagnostic assistance. Functional cholestasis, also known as non-specific cholestasis syndrome, usually develops within the first 2 weeks in the absence of a documented cause [130]. Cholestasis is both hepatocellular and canalicular, affecting predominantly acinar zone 3 and, variably, also zone 2. Associated hepatocyte ballooning may be extensive and occur early [82]. This common type of cholestasis has been attributed to subcellular organelle damage during preservation/reperfusion with subsequent interference of bile secretion. Clinically, the patient presents with increasing jaundice, without fever or other features of acute cholangitis. There is markedly raised conjugated hyperbilirubinaemia and alkaline phosphatase, with only a mild elevation of transaminases. In most instances, the clinical and histological features are completely reversible, but the jaundice may persist for weeks. In some cases, severe hepatocyte ballooning progressed to perivenular cell drop-out, suggesting an ischaemic basis for this lesion.

5 Liver-Allograft Rejection

5.1 Generalities

Rejection can be broadly defined as an immunological reaction to the presence of foreign tissue components that has the potential to result in graft damage, dysfunction and failure [53]. Like other solid organ transplants, rejection of the liver graft is divided into:

- Hyperacute rejection: by definition associated with pre-formed anti-donor cytotoxic antibodies
- Acute rejection: the most common form, in which prime T cells acting on target antigens play a major effector role
- Chronic rejection: defined histologically as obliterative foam-cell arteriopathy and ductopenia, with precise mechanisms that are still uncertain.

When compared with the heart or kidney, the liver has unique characteristics that affect rate, patterns and histological changes of rejection. First, a dual arterial and venous vascular supply, and a well-developed microvasculature – the sinusoids – lined by a vast number of macrophages – the Kupffer cells – make the liver less susceptible to the effects of vascular thrombosis and more able to remove immune complexes, platelet aggregates and fibrin. In addition, the liver seems to release soluble class-I human leucocyte antigens (HLA) [14] capable of mopping up large quantities of antibodies. The distribution of class-I and -II components of the major histocompatibility complex, the main targets of rejection, also plays a determinant role in the rejection pattern. Both are strongly expressed on antigen-presenting cells, which in the liver appear scanty and mainly localised within the connective tissue of the portal tracts. In the steady state, hepatocytes are negative for both class I and class II, whereas bile-duct epithelium and vascular endothelia normally express class I. Aberrant or enhanced expression of HLA molecules under the induction of various cytokines does occur, not only in rejection, but also in various inflammatory conditions affecting the graft [117]. Such a change in expression is bound to have a triggering or modulating effect on the rejection process, an important aspect considering the frequent graft infection by viruses. Finally, the early replacement of highly immunogenic donor Kupffer cells by recipients cells [43] may constitute an additional explanation for the less severe course of liver-graft rejection compared with renal-graft rejection.

5.2 Hyperacute Rejection

This form of antibody-mediated rejection was thought not to affect liver grafts, even when inserted across ABO blood-group barriers. In retrospect, cases of haemorrhagic liver-graft necrosis were likely examples of hyperacute rejection. In the liver, these take days to develop, while in the kidney or in heart grafts, they may develop within hours [115]. Morphologically, the major vessels are patent, the graft is swollen, dark in colour and increased in weight. Histology shows haemorrhagic necrosis, affecting both parenchyma and portal tracts, with fibrin thrombi present in the vessels [46]. Early immunohistochemistry will demonstrate bound immunoglobulin G (IgG), complement Clq and C3, and the diagnosis should be confirmed by the detection of donor-specific antibodies in an eluate from the failed graft [21]. Individual livers might survive grafting across ABO barriers or in the presence of pre-formed IgG lymphocytotoxic antibodies. However, the risk of graft loss within the first 30 days is higher than in matched controls [22].

5.3 Acute Cellular Rejection

5.3.1 Definition and Clinical Features

Acute rejection is defined as inflammation of the graft elicited by a genetic disparity between donor and recipient. It primarily affects interlobular bile ducts, venous endothelia and, occasionally, hepatic artery branches [53]. Acute rejection is the most common cause of graft dysfunction during the first few weeks after transplantation, with a peak incidence between day 5 and day 15. It may, however, recur or present for the first time at any point thereafter, usually in association with inadequate immunosuppression or concomitant viral infection [7]. The reported frequency varies from 50% to 80%, which probably reflects the differences in the minimal criteria used to make a diagnosis, the type of immunosuppression given and the number of biopsies performed [57]. When clinically apparent, there are systemic symptoms with malaise, fever, and a rise or stop in falling of the serum transaminases, γ-glutamyl transpeptidase (GGT) and bilirubin. Leukocytosis and eosinophilia are often present. Unfortunately, clinical and laboratory findings lack sensitivity or specificity, and a liver biopsy or fine-needle liver aspiration is needed to confirm the diagnosis of acute rejection.

5.3.2 Histological Findings

Histological changes are regarded as the "gold standard" for diagnosis, although their respective significance remains unclear. In general, biopsy specimens show various combinations of the classical triad defined by Snover et al. [112], namely mixed portal-tract inflammation, bile-duct damage and venular endotheliitis (Fig. 1).

The portal inflammation varies in intensity from one part of the biopsy specimen to another. Examination of a series of levels is recommended for a semi-quantitative assessment of the process. Lymphocytes predominate and include large forms and immunoblasts. Most of the lymphocytes are $CD8^+$ and, to a lesser extent, $CD4^+$ T cells [71]. In addition, there are fair numbers of neutrophils and eosinophils; the latter said to have a prognostic value [36]. Periportal cell spillover is rare, may be seen after immunosuppression withdrawal and usually reflects a severe form of acute rejection.

Bile-duct damage is variable. The most common change is an infiltration of interlobular bile ducts by both lymphocytes and neutrophils, a vacuolation of the

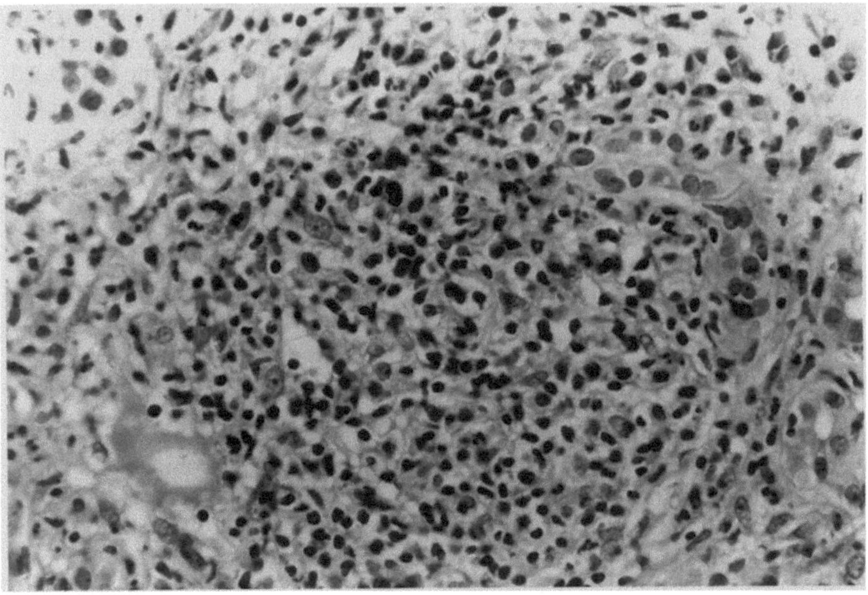

Fig. 1. Portal tract in acute rejection showing a moderate infiltration, predominantly of lymphocytes that obscure the outline of the portal venule (PV) and focally infiltrate bile-duct branches

epithelium with nuclear irregularity, pyknosis and focal disappearance. The dense portal infiltrate may obscure bile-duct branches, but an actual loss of ducts is difficult to ascertain and cyto-keratin staining will often detect duct-epithelium overrun by inflammatory cells [65]. Occasionally, the number of neutrophils in both the wall and lumen of the ducts is such that the lesion mimics a suppurative cholangitis. In this setting, the change seems to reflect severe rejection, rather than ascending cholangitis. At this stage neutrophils are present in large numbers in the bile, which by culture is shown to be sterile [1].

Endotheliitis or venulitis refers to an attachment of lymphocytes to the endothelium that may be lifted up with the formation of a clear subendothelial space in which lymphocytes accumulate (Fig. 2). Endotheliitis affects portal venules, which may be totally obscured by the dense portal infiltrate and, to a lesser extent, hepatic venules where the change is easier to identify due to a lighter surrounding cell infiltrate. Although not a prerequisite, endotheliitis is the most reliable sign of rejection during the early stages. In specimens obtained later, when the range of potential diagnoses is broader, it is of less assistance. On one hand, it is usually less prominent in late cellular rejection, while on the other, a mild degree of endotheliitis occurs in a variety of other conditions, including viral hepatitis. The small arterial branches seen in needle-biopsy specimens may show endothelial swelling and minimal endotheliitis. Fibrinoid necrosis is exceptionally seen in a biopsy specimen. Larger branches appear affected during cellular rejection as revealed by a marked attenuation of medium- and larger-sized arteries on arteriography [24]. Examination of very rare grafts that have failed from acute rejection have reveal-

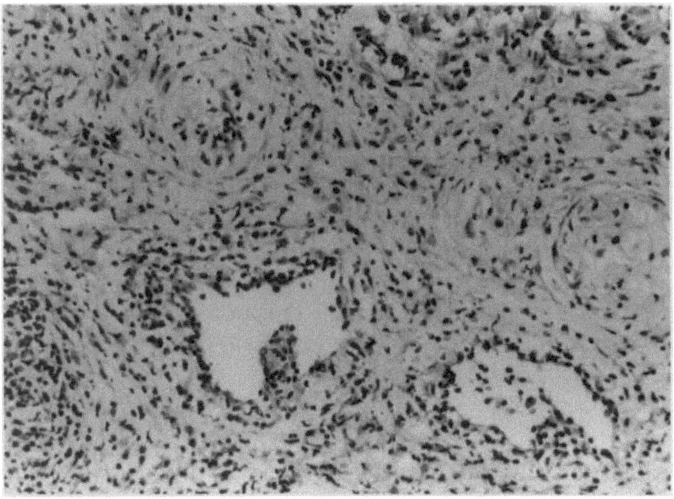

Fig. 2. "Acute vanishing bile duct" combining severe portal venulitis, obliterative foam-cell arteriopathy and absence of identifiable bile duct at 2 months post-transplant

ed prominent luminal narrowing due to intimal inflammation and oedema. Perivenular cholestasis, feathery degeneration, hepatocyte ballooning and loss, and clusters of acidophilic or apoptotic bodies are inconsistently associated with cellular rejection and may reflect the severity of rejection. It is important to note that perivenular changes may be out of proportion or even occur in the absence of significant cellular rejection. In this case, they probably indicate late reperfusion damage and/or non-rejection-related ischaemia.

5.3.3 Grading of Cellular Rejection

The many systems proposed to grade histological changes of acute rejection [23, 29, 55] have been inspired from the one originally proposed by Snover et al. [113]. It is based on scoring the three main histological components of acute rejection. These schemes vary slightly in that an increased emphasis is placed on some features, for example, the numeration of eosinophils [13]. In our centre, for the purpose of trial assessment, we have used the method adopted by the pathologists of the FK506 European multi-centre trial. In short, each of the three elements of the rejection triad are scored on a scale of 0 (none) to 3 (severe), and a final rejection grade of 0–9 is obtained by adding up the scores of these three main features. A final score of 0–2 represents no rejection, 3 borderline, 4–5 mild, 6–7 moderate, and 8–9 severe rejection. Features that are inconsistently seen in acute rejection, such as periportal cell spillover, sinusoidal endotheliitis and perivenular necrosis, are used to upgrade the overall grade of rejection. In practice, we only transmit a global diagnosis to the transplant team, such as no rejection, mild, moderate or severe rejection.

Recently, a panel of recognised experts in liver-transplant pathology have reached agreement on a common nomenclature and a set of histological criteria for the grading of acute liver-allograft rejection [52]. This Banff schema, named after the place where the consensus conference is held, should certainly be adopted widely to improve comparisons of data among various institutions, keeping in mind that inter-observer variations are bound to be significant if pathologists from different centres have not agreed on definitions prior to using the system. However, it must be remembered that the relative importance of the three histological components remains uncertain and, although the qualitative diagnosis of acute graft rejection rests on histology, giving a score to the changes has not always accurately predicted response to treatment, risk of progression to chronic rejection or long-term graft survival. The reader is referred to the original paper for details of the Banff scoring system and additional guidance on biopsy reporting [52].

5.3.4 Outcome of Cellular Rejection: Intractable Rejection

The histopathological diagnosis of acute rejection does not necessarily imply that the process is clinically significant and requires treatment. Whereas severe histological rejection is almost always associated with clinical and biochemical features of rejection, mild and moderate changes have variable clinical correlation. Protocol biopsies have disclosed mild to moderate histological rejection in the presence of normal transaminases. Early episodes of mild rejection can resolve spontaneously without adjuvant anti-rejection therapy [29]. Severity and requirement for treatment are still best defined by clinical features in the context of a biopsy-confirmed rejection.

Following treatment of acute cellular rejection with high-dose steroids, there is a return to normal histology or to minimal residual changes. In repeat biopsies, the mononuclear component of the portal infiltrate has largely disappeared, leaving behind a variable number of neutrophils and macrophages; the bile ducts may appear distorted with an atypical epithelial lining, which includes some piling up of nuclei, probably a feature of regeneration [113]. These post-rejection changes may suggest a cholangitic process and, when pronounced, a possible transition to chronic rejection. This diagnosis would be ruled out by clinical and laboratory recovery and/or histological resolution on a repeat biopsy. The diagnosis of intractable rejection implies the continuation of cellular rejection as evidenced histologically in a patient who has received a standard course of high-dose steroids, and in whom the next treatment change is likely to be a switch in the basic immunosuppression from cyclosporin to FK506 (Prograf).

5.3.5 Differential Diagnosis of Acute Rejection

In most instances, the diagnosis of acute rejection is clear-cut and easily distinguished from changes due to either preservation/ischaemic damage or biliary anastomotic stricture or leak, occurring in isolation. Problems arise when these complications co-exist with rejection.

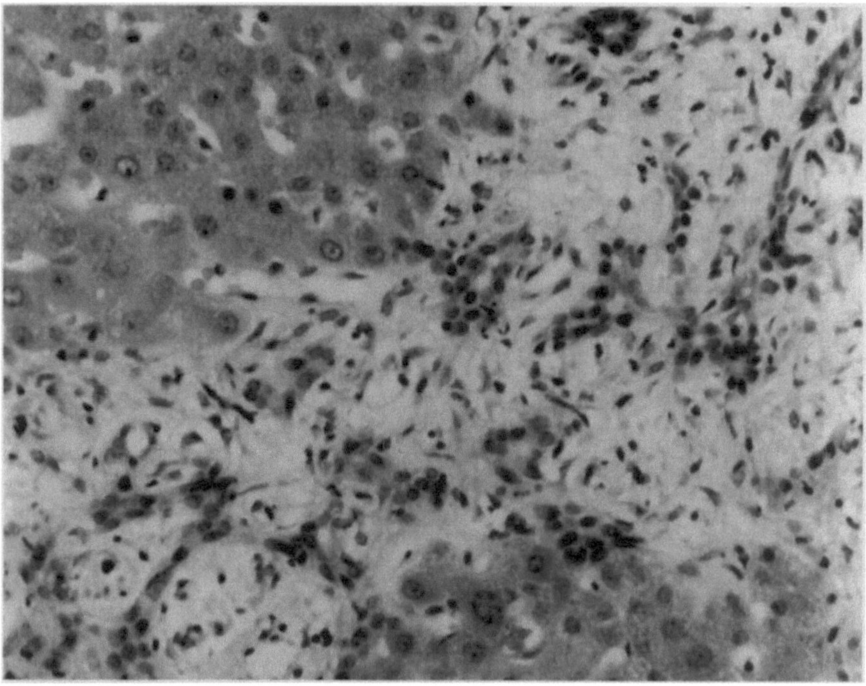

Fig. 3. Enlarged portal tract due to oedema, bile ductular proliferation and sparse inflammatory cells, secondary to a pathological process affecting the major bile ducts

Single or clusters of apoptotic bodies associated with mixed inflammatory cells are frequently observed in the early post-transplant period. During the first 3–4 weeks after surgery, a viral hepatitis does not usually enter the differential diagnosis. Perivenular confluent necrosis and loss with little inflammatory cell reaction is a sign of previous preservation injury or ischaemia. When associated with prominent inflammation, in particular hepatic venular endotheliitis, the change anticipates a more severe form of rejection and may reflect ischaemia due to rejection arteritis.

Portal oedema with a light, mixed inflammatory cell infiltrate and subtle ductular proliferation will favour a biliary pathology (Fig. 3). As mentioned in Sect. 5.3.2, a denser infiltrate including large lymphoid cells and venular endotheliitis, at this stage, almost invariably signifies rejection rather than ascending cholangitis, even in the presence of neutrophils within the wall and lumen of the bile duct.

5.4 Chronic Rejection

5.4.1 Definition and Terminology

If the diagnosis of acute cellular rejection is usually straightforward [23], chronic rejection remains a diagnostic problem due to inconsistency, insidious development and uneven distribution of the changes. Our group was first to add the disappearance of the intrahepatic bile ducts to the long-recognised foam-cell arteriopathy as basic components of chronic liver-allograft rejection, also known as the vanishing bile-duct syndrome [96]. Variable combinations of these two cardinal changes were subsequently accepted as a morphological definition of chronic rejection [37]. Arguments have been advanced against the term "chronic rejection" that imply a time parameter, while the changes as defined histologically can occasionally be present as early as 20 days after surgery – acute vanishing bile-duct syndrome [63]. Synonyms such as "irreversible rejection" or "rejection with duct loss" have been proposed, but the most common term used is "ductopenic rejection", which has not yet supplanted the conventional term of "chronic rejection" [53].

5.4.2 Clinical Presentation

Most cases of chronic rejection present between 2 months and 12 months after transplantation [127]. These follow one to several episodes of acute, but not necessarily severe, rejection in the first 3 weeks. Some have shown a poor response to steroid treatment. Rarely, the presentation is insidious over a period of months, without previously documented episodes of acute rejection. Early on, clinical symptoms resemble those of acute rejection. Later, jaundice develops with a progressive increase in serum GGT, alkaline phosphatase and bilirubin levels. Selective hepatic angiography may show "pruning" of the intrahepatic arteries with poor peripheral filling [24]. Biliary strictures occur in a significant number of patients. The incidence of chronic rejection in different series has varied between 5% and 15% [129], but most centres report a decreasing incidence. Precise reasons for this are not known [81], but occasional protracted cases have been recorded up to 9 years after transplantation [111].

5.4.3 Biopsy Diagnosis

Early changes in a biopsy are difficult to interpret and repeat specimens are often needed to reach a diagnosis. There may be a transitional period when changes of acute cellular rejection persist (intractable rejection), but the cellular infiltration of morphologically abnormal bile ducts gradually lessens as the cholangiole destruction progresses [94] (Fig. 4a–f).

Early changes that favour chronic rejection comprise variable combinations of the following:

a) Light portal inflammation and oedema with damage to the small interlobular bile ducts in the form of distortion, epithelial atypia with nuclear pyknosis and focal disappearance (Fig 5a)
b) Perivenular hepatocyte ballooning and drop-out, with or without early sclerosis and minimal inflammatory reaction (Fig. 5b)
c) Canalicular cholestasis in acinar zone 3 with feathery degeneration and a distinctive absence of periportal ductular reaction (Fig. 4)

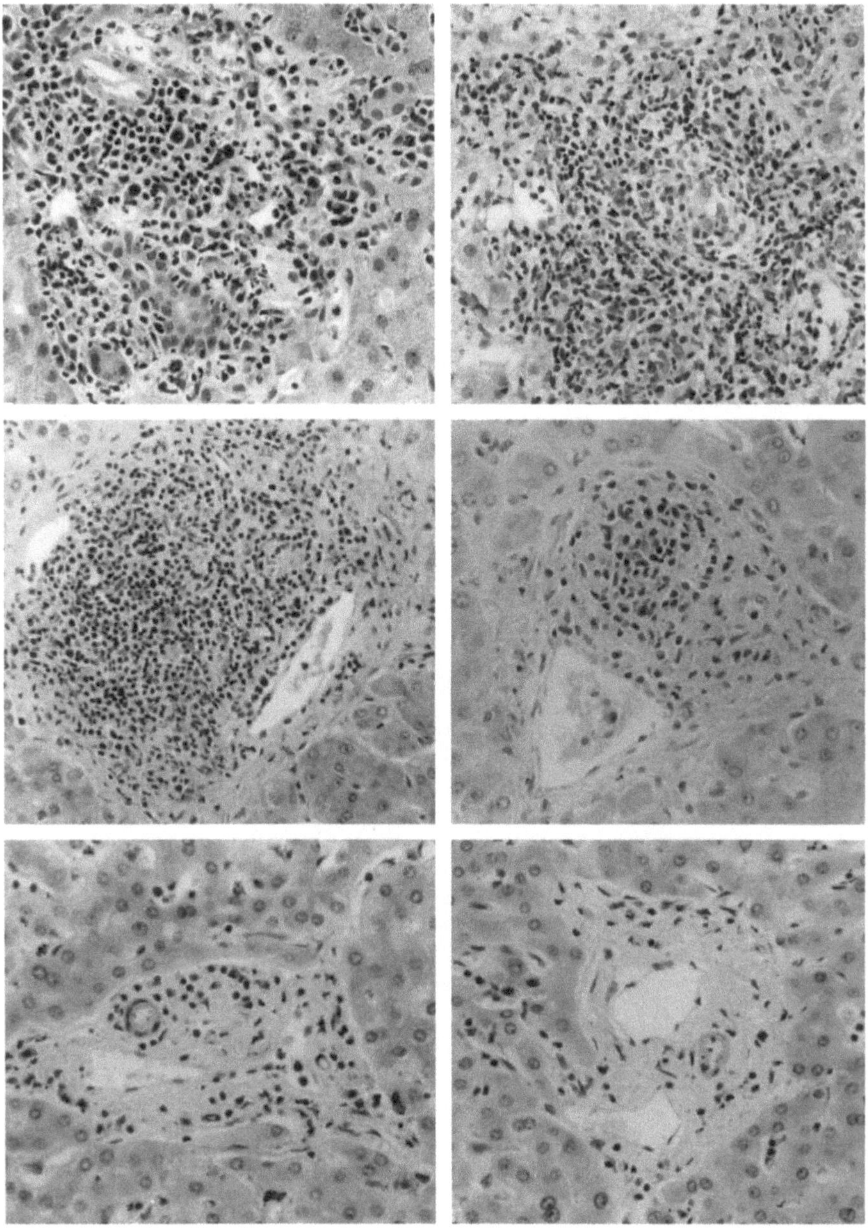
a b c d e f

Fig. 4a–f. Sequential histology in chronic rejection (**a**, **c**);. acute cellular rejection with bile-duct damage (**b**) and loss (**c**); ductopenia with progressive attenuation of the portal cellular infiltrate (**e**, **f**). Note the absence of ductular reaction. **a** 7 days; **b**, **c** 2 weeks and 3 weeks; **d**, **e**, **f** 2 months, 3 months and 7 months after transplant

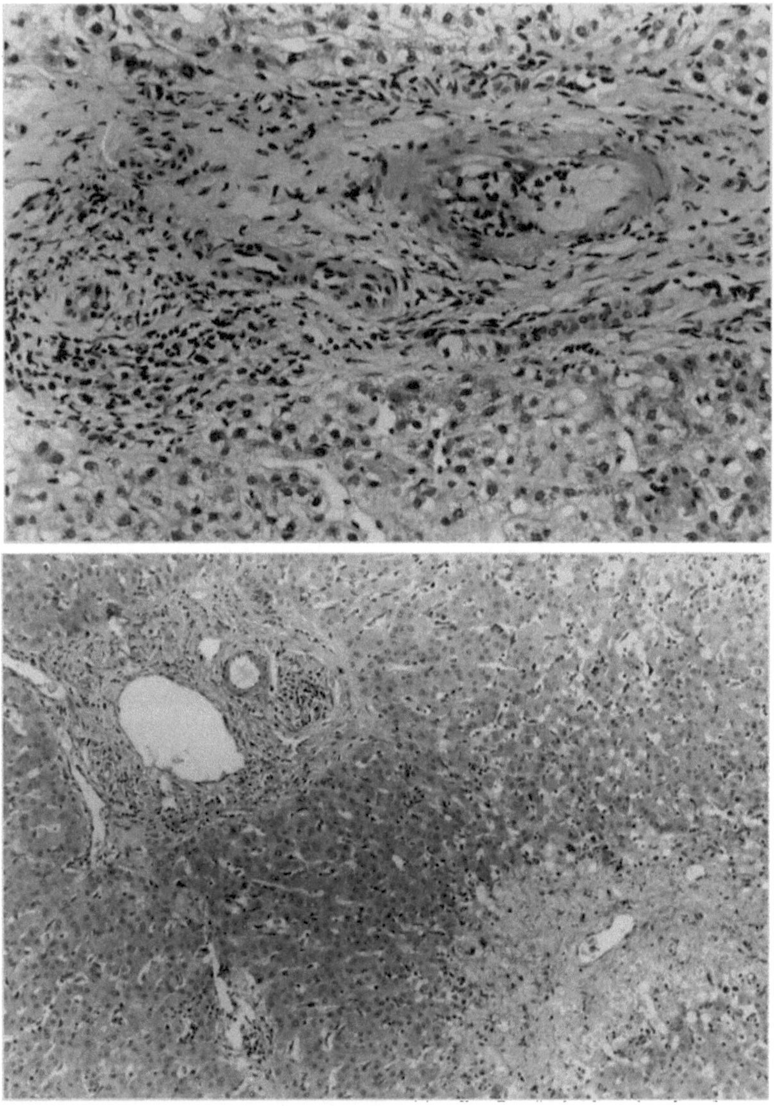

a

b

Fig. 5a–c. Chronic rejection. **a** Portal tract at 2 months post-transplant showing foam-cell endarteritis with an inflamed, yet still identifiable, bile duct to the left of the field. **b** Perivenular confluent cell drop-out and ductopenic portal tract 6 weeks after transplantation. **c** see p. 75

Subsequent biopsy specimens will show a progressive disappearance of the interlobular bile ducts, which affects primarily the small (< 75 µm) branches. Terminal portal axes devoid of bile duct may be difficult to identify as triads. Perivenular distribution of the cholestasis may help in this respect. Estimation of bile-duct loss is achieved by calculating the ratio of portal tracts without identifiable ducts to the total number of portal tracts present. Of the 20 examined, 10 (50%) or more portal tracts devoid of duct have been considered minimal diagnostic criteria [53].

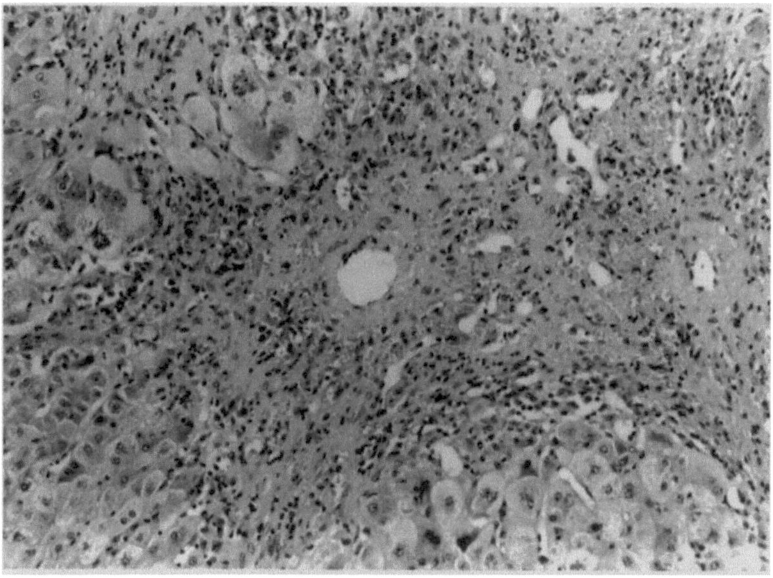

Fig. 5. c Perivenular and bridging cell loss and fibrosis in a graft removed at 8 months after transplantation

Sequential biopsy specimens are often needed over a period of months to reach this figure.

The associated obliterative arteriopathy with intimal deposition of foamy macrophages affects arteries of larger calibre than the ones generally sampled by biopsy needles (Fig. 5a). Its presence is indirectly suspected by persistent perivenular parenchymal cell drop-out. This is supposedly an ischaemic lesion which, when associated with early ductopenia, is a harbinger of evolving chronic rejection [125] (Fig. 5b). Perivenular fibrous-tissue deposition is observed later (Fig 5c).

5.4.4 Explanted Liver

On gross examination, chronically rejected allografts are usually cholestatic and dark green. Slicing will show bright yellow streaks or dots, representing arterial branches that stand out due to intimal thickening and/or luminal occlusion by lipid-rich macrophages (Fig. 6a). On histology, most cases exhibit both foam-cell obliterative arteriopathy (Fig. 6b, c) and severe ductopenia (Fig. 4d–f), but one of the components may predominate and, in occasional cases, seems to occur independently. Ductopenia may be overlooked when restricted to the very small terminal branches, while the larger interlobular and septal bile ducts are well preserved. Conversely, only few muscular arterial branches may show foam-cell deposition. As a consequence, the minimal diagnostic criteria are difficult to establish. The changes in the liver may vary according to fluctuation in pathogenic mechanisms, time of observation after surgery and complicating conditions. In our centre, predominantly vascular cases occur in grafts removed early. Prior with-

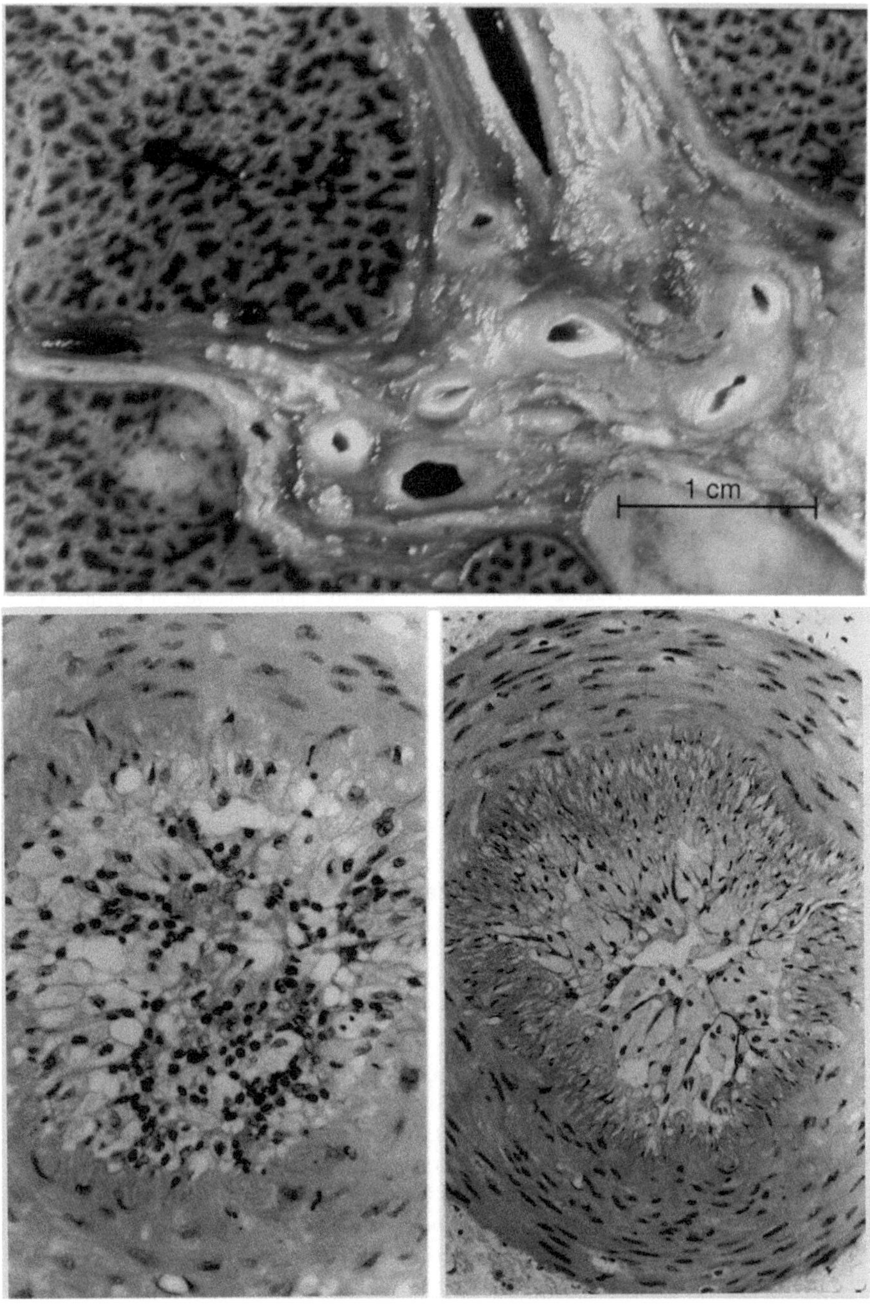

a b c

Fig. 6a–c. Chronic rejection. **a** Bisected liver graft removed at 5 months. Peri-hilar region showing prominent thick-walled artery (*Bright yellow* in colour) and *dark-green* mottling of the parenchyma due to central cholestasis. **b** Early endarteritis with intimal lymphocytes and foam cells obliterating the arterial lumen. **c** Typical foam-cell arteriopathy such as observed later

drawal of immunosuppression is associated with a marked lymphocytic cholangitis affecting septal bile ducts. In some cases, large intrahepatic bile ducts show secretory hyperplasia or metaplasia, pleomorphism of their epithelial lining, ulcerations and strictures, likely due to ischaemic cholangitis. In general, peripheral ductular reaction is minimal or absent. Portal fibrosis is inconspicuous, unless a stricture of a main bile duct had dominated the course. In contrast, perivenular fibrosis (Fig. 5c) is common and may reach the stage of extensive centro-central bridging septa with an apparent reverse lobulation.

5.4.5 Differential Diagnosis of Chronic Rejection

In addition to the difficulty of distinguishing early chronic rejection from unresolved cellular rejection, ischaemic cholangitis (see Sect. 6.2) may produce histological changes in the peripheral parenchyma that mimic those of chronic rejection; in particular persistent perivenular hepatocyte loss and cholestasis. Cholangitic changes with ductular proliferation are, in most instances, more pronounced and ductopenia is usually less severe than in chronic rejection.

5.4.6 Outcome of Chronic Rejection

In the past, chronic rejection was regarded as a progressive and irreversible loss of the intrahepatic bile ducts not responsive to immunosuppressive therapy, which required graft replacement. Since then, clinical and histological improvement, or resolution, has been observed in patients whose graft histology met the criteria of chronic rejection [51, 83]. Furthermore, an early study has suggested that FK506, in addition to being beneficial in cases of intractable rejection, may arrest the progress of ductopenic rejection [20]. Several uncontrolled studies have since reported similar results with graft salvage in 50% or more of chronic-rejection cases. The rate of success presumably depends on the severity of either ductopenia or obliterative vasculopathy at the time of therapy conversion [131]. Histology shows a resolution of canalicular cholestasis, although some degree of ductopenia persists, and in a few specimens a re-colonisation of the original duct channel from marginal bile ductules is reminiscent of duct formation from the embryonic ductal plate.

5.4.7 Pathogenic Mechanisms

The mechanisms of chronic rejection are questionable, and the controversy surrounding the usefulness of HLA matching has not been proven. [28]. T cell-mediated cytotoxicity is thought to take place; a predominance of $CD8^+$ lymphocytes in the portal infiltrate having been associated with a higher risk of destruction of biliary epithelia [71]. In vitro, resting human intrahepatic biliary cells express class-I major histocompatibility complex (MHC) antigens, intercellular adhesion molecule 1 (ICAM-1), and relatively low levels of lymphocyte-function associated antigen (LFA-3), whereas stimulation with interferon-γ alone or in combination with tumour necrosis factor alpha (TNF-α) upregulates class-I MHC antigens and ICAM-1 and induces class-II MHC molecules [61]. Similar changes have appeared on small bile-duct cells during acute rejection.

Overlap or a continuum between acute and chronic rejection is common, and there is some evidence that chronic rejection may, at times, represent the cumulative effect of inadequately controlled acute rejection. The density of the cell infiltrate may diminish as a result of the progressive disappearance of the bile ducts, which are the main target antigens. The arterial lesion, which may have started early in the postoperative course, follows a complex interaction of factors [19], including initial damage to the endothelium, deposition of platelet aggregates and fibrin, altered haemodynamics, increased permeability followed by release of cytokines and growth factors from platelets, endothelial cells or infiltrative immune cells and deposition of serum proteins and lipids. This triggers an influx of macrophages and a repair response that includes myofibroblast proliferation and, later, synthesis and deposition of matrix proteins. Ischaemia secondary to the obliterative arteriopathy may also contribute to the duct damage, as blood supply to the bile ducts is essentially derived from the hepatic artery. The severity of ductopenia parallels that of the obliterative arteriopathy [87]. A morphometric study showed that the microvascular destruction preceded bile-duct loss in both acute and chronic rejection and was most severe in the latter [69]. Persistent cytomegalovirus (CMV) infection as an independent risk factor for chronic rejection [84] has not been widely confirmed.

6 Biliary Complications

Biliary complications were once a major problem, with a reported incidence of 50% and a mortality of 30% in the early series [8, 114]. With improved donor-liver perfusion and surgical techniques, the incidence has been reduced considerably; 7%–29% in more recent reports, with a mortality of less than 5% [42]. Liver biopsy has relatively high sensitivity in spotting a biliary pathology as the likely cause of graft dysfunction, but it is of little value compared with radiological techniques in determining the type and site of biliary injury.

6.1 Biopsy Diagnosis

As in the non-transplant liver, changes secondary to a biliary pathology (Fig. 3) are characterised by portal oedema, a marginal ductular proliferation and a predominantly neutrophilic and eosinophilic polymorph infiltrate that varies from sparse cells to dense clusters often found within the wall or lumen of the ducts. A variable degree of perivenular cholestasis is present. Later in the post-transplant course, histology may reveal chronic biliary features in patients with an unsuspected biliary complication. In this situation, there is portal or portal-septal expansion due to loose connective tissue, ductular proliferation and a biliary type of limiting plate disruption that sometimes includes cholate-static changes with copper-asso-

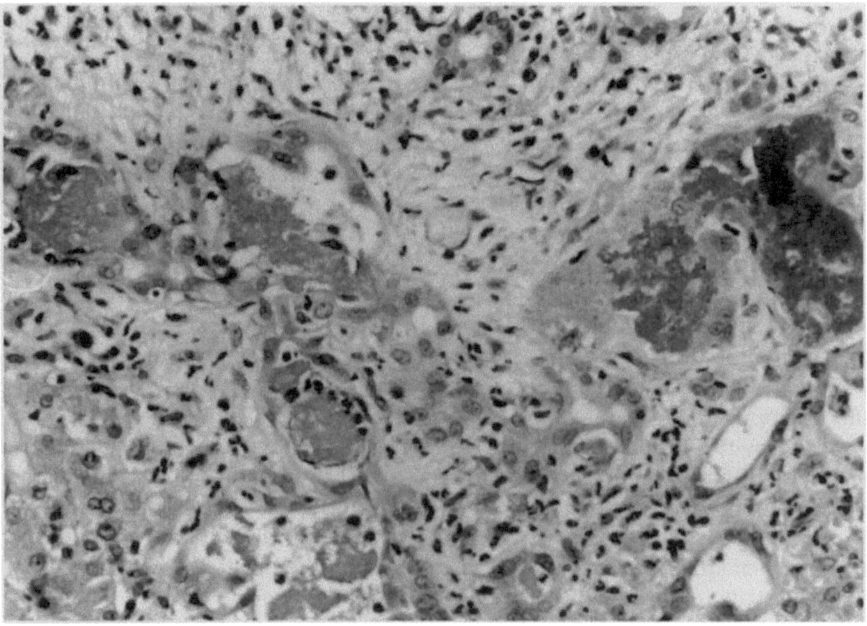

Fig. 7. Biliary sepsis. Oedematous portal tracts with dilated cholangioles, which are filled with inspissated bile, some having lost their epithelial lining and being surrounded by neutrophil polymorphs

ciated protein deposition. Such a picture on an allograft biopsy often leads to an endoscopic retrograde cholangio-pancreatography (ERCP) or a percutaneous cholangiogram, depending on whether the patient had a direct duct-to-duct or a duct-to-Roux loop biliary anastomosis.

When histology shows concomitant perivenular hepatocyte damage and drop-out, the biliary injury is likely to have an ischaemic cause, otherwise known as ischaemic cholangitis [64]. Superimposed changes of "suppurative cholangitis", with numerous neutrophils overrunning small interlobular bile ducts, do not necessarily mean infection. Leakage of bile component may produce chemical cholangitis, a likely explanation during rejection cholangitis. The formation of microabscesses, pus in the lumen of larger ducts and inspissated bile casts in periportal-dilated ductules most often reflect septic complications. Ductular cholestasis, as seen in the non-transplant patient, is often a marker of generalised infection or septicaemia, whether this has originated in the biliary tree or not (Fig. 7).

6.2 Types and Timing of Biliary Complications

Bile leaks, usually at the site of the biliary anastomosis or at the insertion site when a T-tube has been used, occur early after transplantation or at the time of T-tube removal. They can often be repaired endoscopically; but those developed at the site of a duct-to-Roux loop anastomosis usually require surgical re-fashion. The T-tube itself may be the cause of the obstruction in some cases.

Anastomotic stricture usually affects duct-to-duct anastomoses within the first 2–6 months and presents as an asymptomatic rise in serum alkaline phosphatase and GGT activity, or with histological evidence of large duct obstruction. Anastomotic strictures, as well as hilar strictures at the bifurcation of left and right hepatic ducts, may have an ischaemic cause and be presenting features of hepatic arterial thrombosis or chronic rejection.

Multiple non-anastomotic strictures [62] are now well recognized as serious biliary complications that often require liver replacement. The clinical presentation is similar to that of the anastomotic complications. However, this occurs later after transplantation, and signs and symptoms of acute cholangitis may be found. Radiologically, the changes mimic those of a sclerosing cholangitis with dilatations and strictures of the intrahepatic bile ducts. This type of complication has been associated with ABO blood group incompatibility [108] and ischaemic cholangitis, whether it is related to thrombosis or to foam-cell arteriopathy [105]. An increased incidence of diffuse extra-anastomotic strictures has also been reported in patients whose initial diagnosis was sclerosing cholangitis [68]. Finally, the complication has also been attributed to a prolonged cold ischaemia time [62]. In this situation, biliary destruction is thought to be a late result of the damage inflicted to the duct lining by stagnating bile during the period of cool ischaemia. It has been suggested that better technique of donor-liver harvesting, including extensive flushing of the biliary tree with the preservation solution, reduces this type of complication.

Biliary strictures often remain asymptomatic for a considerable period of time. Advanced stage-3 biliary fibrosis and occasionally cirrhosis detected on biopsy might be the first manifestation years after transplantation. This was common in patients who had a biliary conduit as a first biliary anastomosis, a technique which, although preventing early biliary strictures, has been followed by progressive bile sludging. This is now of historical interest as the technique is no longer performed, and the majority of the patients who had this procedure prior to 1989 have had their biliary conduit subsequently replaced by a duct-to-Roux loop anastomosis. In rare cases, extrinsic bile-duct compression by exuberant amputation neuromas has been the cause of late large-duct obstruction.

6.3 Allografts Removed due to Biliary Complications

Allografts removed during the first 6 months after transplantation show bile-duct ulceration, necrosis, dilatations and/or strictures. These are often associated with areas of infarction. In some grafts, diffuse biliary-tract destruction is characterised by a dark-green staining of the tissue surrounding intrahepatic bile ducts, a pattern which is in keeping with a diffuse destruction of the duct lining with consequent bile extravasation. There may be large cavities with ragged margins, which sometimes contain bile-stained pus, necrotic debris, bile sludges or stones. Histology confirms ducts ulcers or destruction with a surrounding layer of bile-stained ghost tissue or inflamed granulation tissue lining irregularly dilated cavities (Fig. 8). There may be colonies of bacteria or fungi, the growth of which is facilitated by

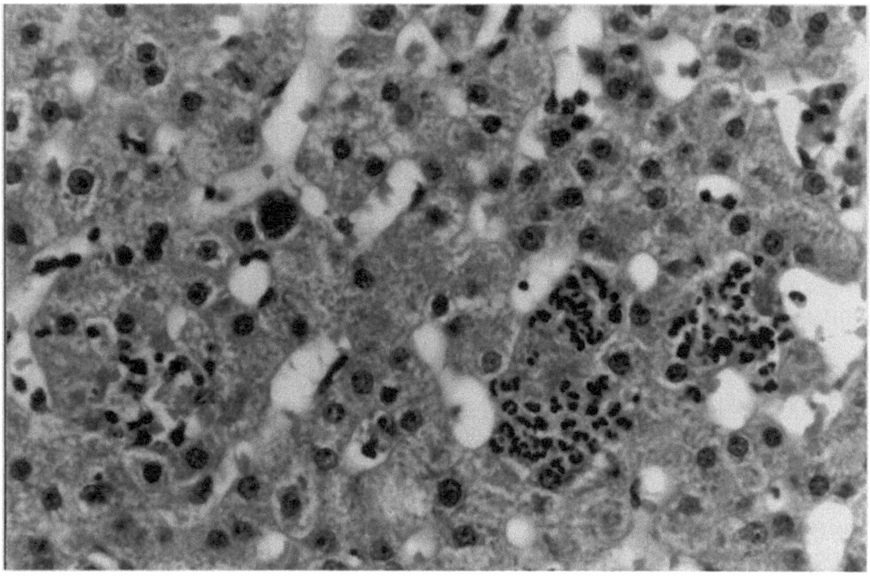

Fig. 8. Ischaemic cholangitis. Necrotic duct wall (lumen at the *lower part* of the field) showing ghost tissue with extravasated bile and a layer of fungi (*black filaments*)

the reduced blood supply, which is likely to interfere with antibiotic's delivery. Bile sludges reveal a collagen skeleton, suggesting their origin from sloughed-off portions of the necrotic bile-duct wall that acted as a nidus. In general, the changes seen in removed grafts affect the perihilar regions predominantly and unevenly, and are much more extensive than what might have been expected from the appearance of the biopsy.

7 Infection

Liver-transplant recipients are particularly prone to develop the types of infections related to major hepato-biliary surgery and to the use of intravascular catheters, as well as the spectrum of opportunistic infections known to arise as a consequence of immunosuppression. In addition, the liver allograft is the prime target of de novo or recurrent infection by the hepatotropic viruses.

7.1 Viral Infections

7.1.1 De Novo Hepatitis B and C Infection

Routine screening for HBV and, since 1992, for HCV has drastically reduced HBV and HCV de novo infections. Exceptional cases of acute HBV infection are reported from hepatitis B surface antigen (HBsAg)-negative donors who were carrying anti-HBc antibodies. In our centre, the rate of acquisition of HCV infection has fallen from 3.7% to 0.3% since the introduction of routine and sensitive testing of both blood and organ donors [40]. Nevertheless, hepatitis B and C have to be considered in the differential diagnosis of a predominantly lobular hepatitis, found in biopsy specimens taken from 4 weeks onwards. At times, virus-like hepatitis changes are observed in the absence of any serum markers and the cause remains undetermined. The newly described hepatitis G virus (HGV), whether co-transplanted with HCV or acquired, seems to have an impact neither on the clinical course nor on the allograft pathology [38]. Thus, the majority of cases serologically confirmed as HBV or HCV infection represent reinfection and are considered later.

7.1.2 Cytomegalovirus

CMV infection, which has a peak incidence at about 25–38 days, may take the form of a non-specific hepatitis. CMV infection is suspected when dense clusters of neutrophils (microabscesses) are found throughout the parenchyma. Multi-step sections may reveal pathognomonic owl's eyes inclusions (Fig. 9). These are uncommon and focally distributed, particularly in specimens taken early in the course of the disease when immunostaining considerably increases the diagnostic yield [90]. In this respect, it is essential to use an antibody directed at an early CMV antigen (MAB 810, Chemicon), which detects nuclear CMV long before typical inclu-

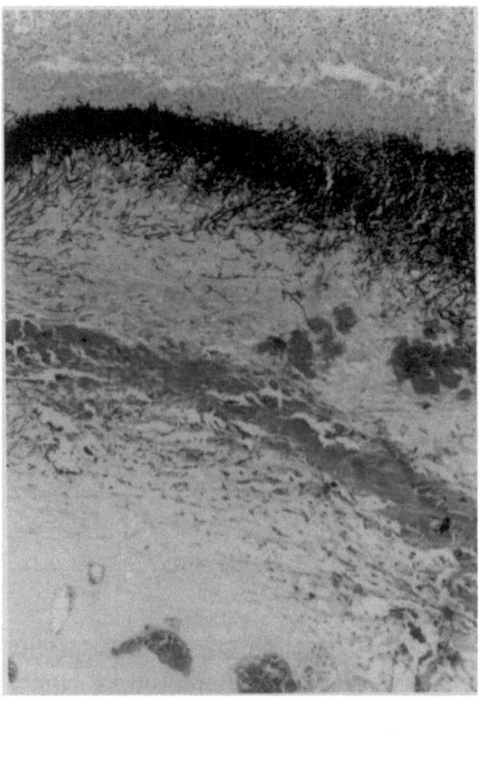

Fig. 9. Cytomegalovirus infection characterized by microabscesses and a single large nuclear inclusion

sions can be identified. We find this technique more sensitive than the much more demanding, yet highly specific, in situ hybridisation [78]. Since the advent of prophylactic regimens, histological detection of CMV infection in graft biopsy has become rare. Microabscesses can be seen in CMV-unrelated cholangitis. Recently, numerous mini-microabscesses were found in the biopsies of patients in whom a cause has not been identified; these appear to have no adverse effect on graft and patient survival [66].

7.1.3 Epstein Barr Virus

Both primary and secondary infections with Epstein Barr Virus (EBV) have been associated with acute and chronic graft disease. Histologically, a lobular hepatitis with a characteristically diffuse infiltration of sinusoids by lymphocytes may occur, similar to that seen in non-immunocompromized hosts. In some patients, EBV hepatitis is associated with a dense portal infiltrate of large lymphocytes that can be difficult to distinguish from that of acute cellular rejection. The predominance of B cells, the absence of eosinophils, and a more widespread sinusoidal infiltration favour EBV infection. The diagnosis is confirmed by detection of viral antigen on immunostaining, using monoclonal antibodies to EB nuclear-associated antigen or by in-situ hybridization [119]. The subsequent development of lymphomas in a small proportion of patients is considered later in the chapter.

7.1.4 Rare Opportunistic Viruses

Herpes simplex virus is a rare cause of allograft hepatitis, usually affecting children. It is characterised by foci of parenchymal necrosis, containing neutrophils and macrophages that may develop into extensive areas of confluent necrosis. Similar changes have occasionally been observed with varicella zoster virus (VZV) [18]. Dense intranuclear inclusions are seen in viable cells at the edge of the necrotic areas, but they are absent in biopsy specimens taken early. Adenovirus may produce a similar picture, particularly in children who required additional immunosuppression [74]. Early changes resemble the microabscesses seen in CMV hepatitis. Immunochemical staining, using an adenovirus-group antiserum (MAB 805, Chemicon) allows an early diagnosis [106].

7.2 Bacterial Infection

Although bacterial infections are almost invariably present after liver transplantation, morphological evidence of graft involvement with later abscess formation usually occurs in the context of biliary tract complications. The bile usually becomes colonised between days 7 and 10 and any interference with bile flow will make colonisation of the graft easy. Consequences are dramatic when ischaemia is associated with ischaemic cholangitis. Infarcted tissue becomes an ideal culture medium inaccessible to antibiotics and progressive suppurative destruction of bile ducts and surrounding structures rapidly leads to abscess formation, usually an indication for urgent re-transplantation. Ischaemic cholangitis affecting the larger ducts may be observed on biopsy in the form of pus detected in small bile-duct lumina, but this is not always the case. Persistent cholestasis, particularly if ductular bile casts are present, and subtle cholangiolitic changes may be the only sign of severe biliary sepsis (Fig. 7).

7.3 Fungal Infection

Both *Candida* and *Aspergillus* species may give rise to systemic infection, but these organisms are rarely observed in needle biopsies. They may be found in the necrotic areas of allografts removed for biliary sepsis (Fig. 9). Fungal infection has occasionally resulted in pseudoaneurysm formation and rupture of the hepatic artery. Other opportunistic infections, such as *Pneumocystis carinii*, and isolated cases of listeriosis and cryptococcosis have been reported affecting extrahepatic sites.

8 Disease Recurrence

The possibility of disease recurrence after transplantation must be part of the continued evaluation of all liver-allograft recipients. This may have a significant impact on both the quality and the length of patient survival and, in this respect, may influence future patient selection for transplantation. The natural history of disease following transplantation is likely to be modified by the administration of immunosuppression; therefore, criteria for the diagnosis of recurrent disease may differ from those used to diagnose the original disease [96].

8.1 Hepatitis B Virus

Following the first report of HBV recurrence after liver grafting [124], it soon became apparent that, without preventive measures, the vast majority of patients with circulating HBsAg at the time of surgery would reinfect their graft [17]. The resulting reduction in both graft and patient survival was important enough to question the role of liver transplantation in the management of HBV-induced liver disease. Since then, prophylactic measures have changed the outcome and good results are now achieved in this group of patients [103].

8.1.1 Natural History of Hepatitis B Virus Post-transplant Reinfection

Early series showed the importance of viral replication at the time of surgery in determining the risk of recurrence in patients who had received minimal or no immunoprophylaxis. In a survey of 334 patients from 17 European centres, reinfection defined by HBsAg in serum was highest (67%) in patients with HBV-related cirrhosis, and lowest (17%) in patents with fulminant hepatitis B (17%). The risk of reinfection in patients with hepatitis D virus (HDV) co-infection was intermediate; 40% for those with a fulminant hepatitis and 32% for those with cirrhosis [104]. Within the HBV-related cirrhotic subgroup, the risk of recurrence was higher in those seropositive for HBeAg and HBV DNA than in those seronegative for both (83% vs 58%). This high recurrence before the use of immunoprophylaxis severely affected survivals with a 2-year actuarial survival rate of 51.7% in HBV-positive patients compared with 74.4%–88.0% in patients with various chronic liver disease [86].

8.1.2 Pathology of Hepatitis B Virus Graft Reinfection

The first histological evidence of HBV recurrence is the demonstration of nuclear and cytoplasmic hepatitis B core antigen (HBcAg), approximately 2–5 weeks post-transplant [92]; later, cytoplasmic and membranous HBsAg is present. A mild lobular hepatitis does not ensue before 8–10 weeks and usually corresponds to the first graft dysfunction, which can be attributed to HBV infection [86]. The subsequent course cannot be anticipated at this stage; histological changes recapitulate the full spectrum of liver lesions seen with naturally occurring HBV infection, but the disease progresses rapidly and a significant proportion of the patients have cirrhosis within an average of 2.5 years after surgery [123]. In addition, fibrosing cholestatic hepatitis (FCH) [15] develops in approximately 25% of reinfected grafts, with a rapid progression to graft failure. FCH is characterised by widespread hepatocyte ballooning, vacuolation and loss, and comparatively mild inflammatory cell reaction. CK19-positive neo-ductules extend from the periportal areas into the acinus accompanied by extensive perisinusoidal fibrosis, and cellular and canalicular cholestasis is often severe (Fig. 10 a, b). There is a massive accumulation of both HBV genome – shown as diffuse HBV-DNA staining by in-situ hybridization – and HBsAg and HBcAg products. Other groups have emphasised different aspects of the lesion. Benner et al. [5] use the term fibrosing cytolytic liver failure to explain what might be the mode of cell death. Phillips et al. [92] distinguish a form with particularly extensive fatty infiltration – steato-viral hepatitis B. In spite of these differences, all studies report a rapidly progressive fibrosing lesion with little inflammatory reaction and an extreme degree of viral load, quite unlike any pattern previously described in HBV infection. Once established, the lesion appears refractory to therapeutic measures and re-transplantation has resulted in an accelerated pace of disease recurrence and progression [86].

8.1.3 Mechanism of Reinfection and Pathogenesis

HBV recurrence may result from immediate reinfection during transplantation or soon after. Most likely, it recurs from the extrahepatic reservoir, particularly in late reinfection. HBV has been detected, and its replication demonstrated in peripheral-blood monuclear cells after transplantation, prior to its reappearance in the

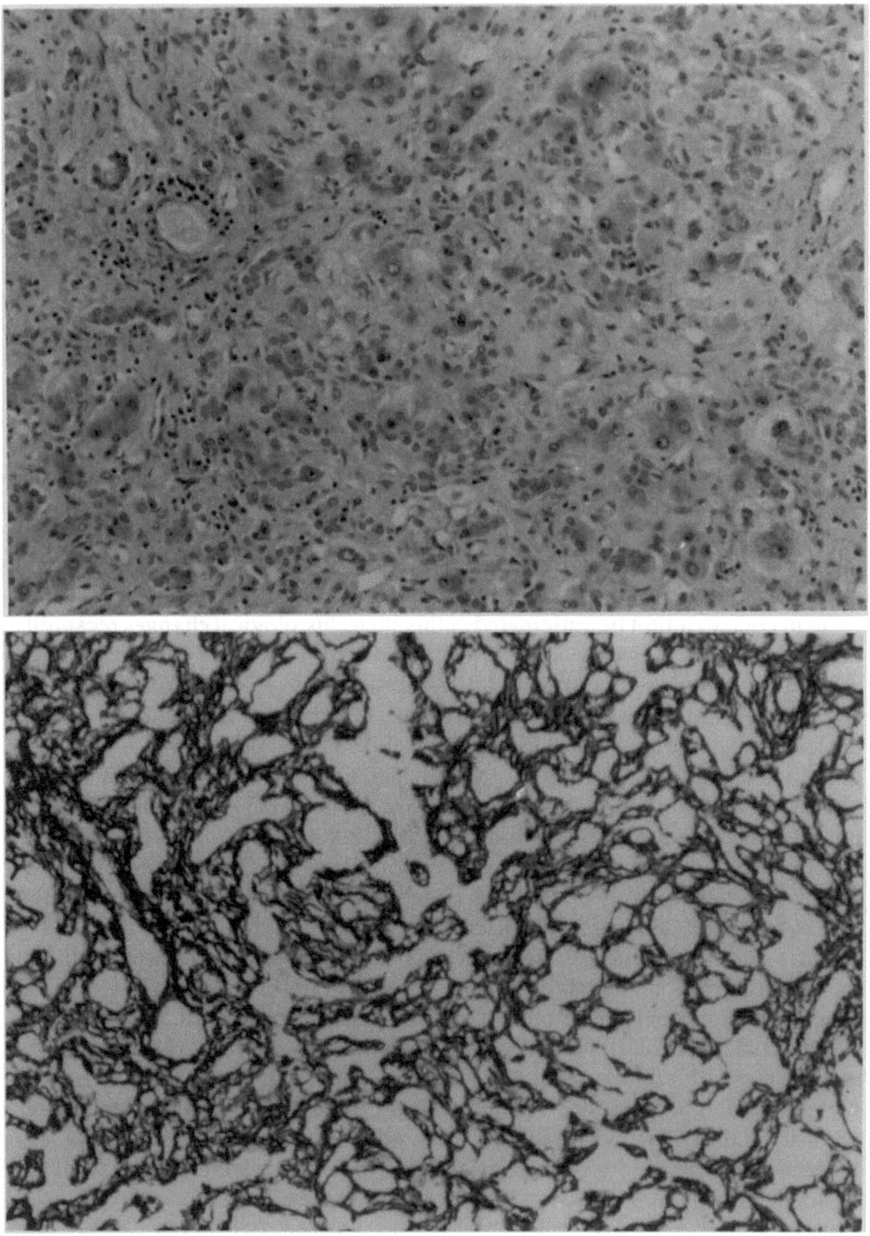

a

b

Fig. 10 a, b. Fibrosing cholestatic hepatitis due to recurrent herpes B virus (HBV) in the graft. Disruption of the parenchyma with extensive pericellular fibrosis and intermingled neoductules; **a** H and E; **b** reticulin

graft [32]. Immunosuppression may be responsible for the rapid progression of the liver changes in some patients. Additional factors might explain the unique patterns observed after liver grafting. HBV-induced liver damage is mediated via HLA cytotoxic T lymphocytes directed at HBc-peptides expressed in association with identical class-I HLA antigens at the hepatocyte membrane [76]. The lack of HLA identity between graft and recipient might, therefore, interfere with the effectiveness of the cytotoxic T-cell response and contribute to the unprecedented accumulation of viral material. This was the case in one FCH liver with 10^{18} core particles [92]. A direct cytopathic effect of HBV has been shown in transgenic mice in which an overproduction of pre-S1 inhibits HBsAg secretion from the endoplasmic reticulum [10]. A parallel has been drawn between that experimental model and human FCH, in which the massive viral burden might similarly interfere with cell function, leading to severe hepatocyte damage with minimal inflammatory reaction. Specific viral mutations in the pre-core region has also been linked with severe post-transplant HBV reinfection [2].

8.1.4 Hepatitis D Virus Co-infection

As a subgroup, patients chronically infected with both HBV and HDV have a lower risk of recurrence than patients infected with HBV alone (50% vs 80%) [88]. It is likely that the majority of patients with post-hepatitis HDV cirrhosis are HBV-DNA negative at the time of transplantation. It is also possible that HDV exerts a repression on HBV replication and diminishes the risk of transmitting the helper HBV necessary for full HDV reinfection. Interestingly, HDV antigen may be seen in the majority of grafts within a few days without detectable HBV genome or gene products and in absence of virus-induced liver damage [16]. Later recurrence of HBV antigens in these livers at 3–6 months is associated with a massive expression of HDV antigen and an overt lobular hepatitis, suggesting that HDV requires HBV to cause liver damage.

8.1.5 Effect of Prophylaxis and Treatment

Early experiences determined that the main prophylactic measures are patient selection and passive immunoprophylaxis. Thus, from the early 1990s, patients with fulminant hepatitis B or D, chronic HDV infection or chronic HBV infection without detectable HBeAg or HBV DNA are primarily considered for transplantation. Immunoprophylaxis is achieved by repeated long-term administration of HBsIg, started during the anhepatic phase. With such measures, the rate of recurrence has been reduced from approximately 70% down to 30% after 3 years, and survival rates are now comparable to those obtained in elective HBV-negative recipients [25, 104]. HBsIg has limited availability and is very costly. Many patients show evidence of viral replication prior to being considered for orthotopic liver transplantation (OLT). Therefore, alternative therapies are being tested, in particular, interferon-α and nucleoside analogues, such as lamivudine and famciclovir. These compounds are variably evaluated in trials both as pre-emptive treatment to prevent graft reinfection and as post-transplant treatment for patients with overt

HBV recurrence. Preliminary results are encouraging, but also record some treatment failure. Mutations in the HBV genome that facilitate the "escape" of HBV from the inhibitory effects of the specific treatment are a likely explanation in some cases and anticipate the future need to use combined therapies [120].

8.2 Hepatitis C Virus

8.2.1 Pre-transplant Hepatitis C Virus Infection

The possibility to test for antibodies to HCV recombinant antigens has revealed a high prevalence of HCV infection among patients who underwent transplantation. Retrospective studies show that approximately 13% – 16% of recipients had serologic evidence of HCV infection [99]. This affects not only patients with a clinical diagnosis of hepatitis C, but also one third of the patients with end-stage alcoholic liver disease, 10% – 27% of those with chronic HBV infection and at least half of the patients with cryptogenic cirrhosis. Patients with end-stage primary biliary cirrhosis (PBC) and sclerosing cholangitis are rarely positive. Interestingly, HCV is not detected in either the blood or liver of patients transplanted for non-A, non-B-acute or late-onset liver failure [102].

8.2.2 Incidence and Impact of Hepatitis C Virus Reinfection

The incidence of HCV recurrence was grossly underestimated in early studies due to a low rate and late development of antibodies in the immunocompromized host [109]. Using polymerase chain reaction (PCR), HCV RNA was detected in serum and/or liver tissue of 72% and 95% of patients whose liver and/or serum had been positive for anti-HCV antibodies, second-generation recombinant immunoblot assay (RIBA-2) or HCV RNA, respectively [33]. Now, most centres agree that recurrence detected by PCR is almost universal.

The consequences of HCV infection on the liver allograft are less devastating than those observed following HBV infection. Post-transplantation hepatitis occurs in approximately 40% of patients with HCV reinfection [132]. Although the disease seems to run an accelerated course when compared with its counterpart in the non-transplant patient, the impact on the 5-year survival rate appears negligible [40]. Genotype 1b and higher levels of viraemia post-transplantation have been linked with a more aggressive course [33, 41], although high levels of HCV RNA can be found in the absence of allograft damage [9].

8.2.3 Pathology of Hepatitis C Virus Allograft Infection

There are no significant clinical or pathological differences between acquired or recurrent HCV infections [34]. When recognized histologically, HCV recurrence appears as a mild portal and lobular hepatitis with plate disarray, acidophilic bodies and a mild and inconstant fatty infiltration. This may be accompanied by a

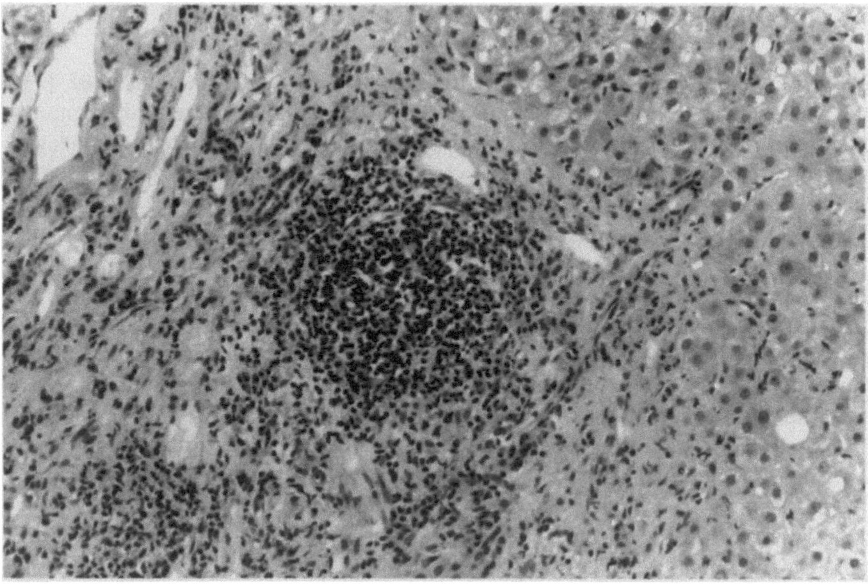

Fig. 11. Hepatitis C recurrence in liver graft. Biopsy at 8 months, showing an enlarged fibrotic portal tract with a predominantly lymphocytic infiltrate including one aggregate and mild interface hepatitis

rise in transaminase activity, but usually does not occur before 6 weeks. Progression to CH with portal lymphocytic aggregates occurs in 50% of the patients after 1 year and 79% after 4 years [45] (Fig. 11). Bile-duct injury is occasionally found [121]. Of 130 patients surviving more than 6 months, 12% had no evidence of CH up to a median of 20 months follow-up; 54% had mild CH; 27% moderate CH within a median follow up of 35 months; whereas ten patients had documented cirrhosis at between 2 years and 11 years post-transplant [40]. Rare cases of progressive cholestatic hepatitis leading to allograft failure within the first 2 years after surgery have been ascribed to recurrent hepatitis C [107].

The changes of hepatitis C may be difficult to distinguish from those of acute cellular rejection, which at times may co-exist. In HCV infection, the portal infiltrate is predominantly composed of mature lymphocytes. These tend to form aggregates and show an intimate association with small bile ducts, which may be infiltrated, but not destroyed, and with portal venules with no frank endotheliitis. At this late stage post-OLT, the associated yet mild lobular hepatitis also favours HCV infection. In most cases where there is doubt, the changes are overall mild, and a clinical evaluation of HCV-RNA results and the level of immunosuppression may be of assistance. It is important to remember that untreated mild rejection is often innocuous [29].

8.3 Primary Biliary Cirrhosis

The possible recurrence of PBC in the allograft, first suggested in 1982 [79], has been the subject of considerable controversy in the literature. This is mainly due to the lack of a specific marker. The persistence after transplantation of elevated IgM and anti-mitochondrial antibody (AMA), as reported to a variable extent in most studies [30]), and the recurrence or de novo development of extrahepatic disorders commonly associated with PBC suggest that the underlying host defect remains uncorrected by removal of the diseased liver; it does not necessarily follow that disease is affecting the graft. Histology remains the "gold standard" by which disease recurrence is identified. However, even histology may be difficult to interpret since the rejection process is focused on the bile ducts, and stringent criteria, such as the presence of granulomas, are required (Fig. 12a), a requirement which is bound to underestimate the actual rate of recurrent disease.

Pathologists from different centres have now observed convincing granulomatous bile-duct destruction on biopsy specimens from patients who received a graft for end-stage PBC, 3–4 years earlier [4]. In a large study, the incidence of recurrence was 16% and most of the reported cases were classified as grade 1–2, and sometimes grade 3 [93]. Other groups argue against PBC recurrence [44]. The apparent discrepancy between studies may reflect differences not only in the diagnostic criteria used, but also in the size and frequency of biopsy specimens, considering the patchy distribution of the lesion in its early stage. In addition, an early recurrence observed in patients on low immunosuppression [131] and the fact that the majority of the patients in centres having observed recurrence were on low-dose or no steroids at that time, suggest that immunosuppression regimens do play a role [27]. Also, a recent study from a centre where recurrence was not a feature in the past documents disease recurrence in two of 13 PBC patients following weaning of immunosuppression [70]. Interestingly, a re-expression of pyruvate dehydrogenase complex (PDC-E2), the main AMA component, has been shown recently at the apical region of the bile-duct epithelium in 28 grafts of 38 PBC patients, but in none of 29 controls, and 8 of the 28 grafts showed concomitant histological changes of PBC recurrence [126]. It must be noted that disease recurrence seems to have little impact on either the quality of the patient's life or medium-term survival after transplantation, and the potential recurrence is at present more of academic than clinical importance [95].

8.4 Primary Sclerosing Cholangitis

In comparison with PBC, the diagnosis of possible recurrence of primary sclerosing cholangitis (PSC) is more complex. In the patient's native liver, the diagnosis of PSC is made primarily by imaging the biliary tree, with demonstration of beading, strictures and dilatation of intra- and/or extrahepatic bile ducts. Intrahepatic cholangiographic appearances and histological features of PSC are usually non-specific. Serologically, the presence of anti-neutrophil cytoplasmic antibodies (ANCA) has a variable incidence with the disease and, just as with PBC, the persistence of ANCA post-transplant does not imply recurrence of the disease in the allograft.

a

b

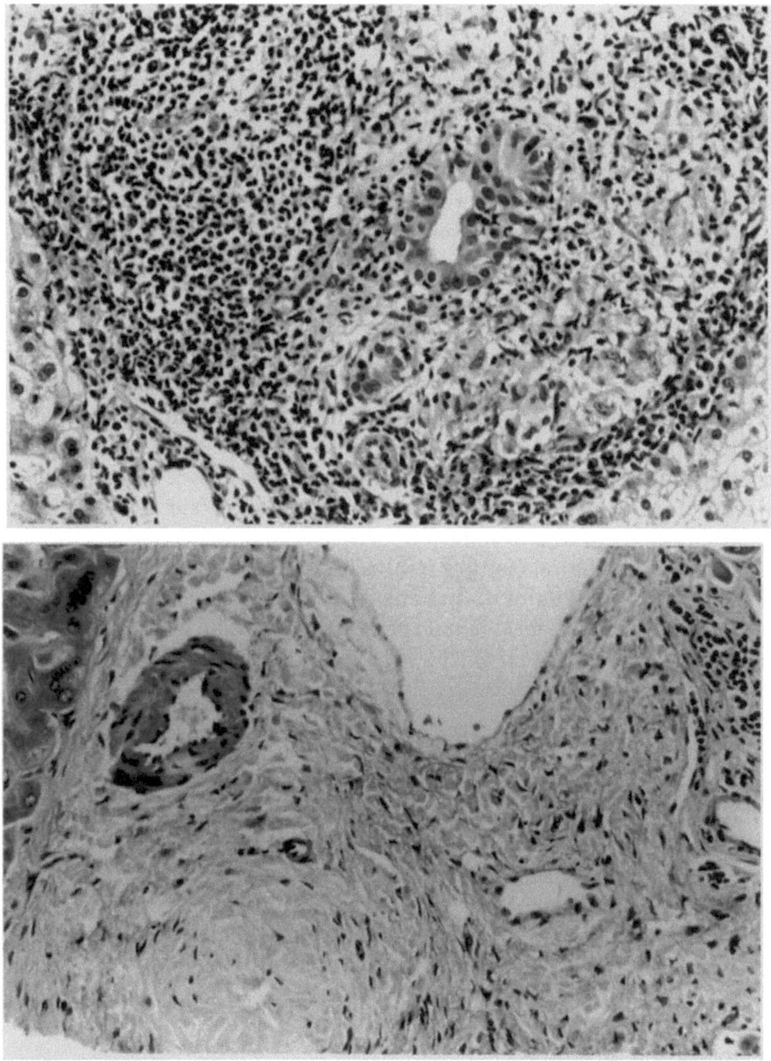

Fig. 12. **a** Recurrent non-suppurative and granulomatous cholangitis in a graft inserted 3.5 years earlier for end-stage primary biliary cirrhosis. **b** Fibro-obliterative duct lesion (*lower left* of the field) 2 years after transplantation for biliary cirrhosis due to primary sclerosing cholangitis

An increased incidence of extra-anastomotic biliary strictures has been recorded in patients transplanted for PSC. However, most patients transplanted for PSC will have a duct-to-Roux loop anastomosis, which may have a higher incidence of biliary complications than a duct-to-duct anastomosis. As already discussed, a similar pattern of non-anastomotic biliary strictures and beading can also be produced by ischaemic-related biliary complications, which may follow thrombotic arterial occlusion, extended cold ischaemic graft preservation or immunologic injury (ABO incompatibility, rejection arteriopathy) [108].

Nevertheless, in a large retrospective study, a "blinded" analysis of post-transplant biopsies showed a higher incidence of biliary complications, including characteristic fibro-obliterative duct lesions, in patients transplanted for PSC than in other disease groups [48] (Fig. 12b). These data add to the growing evidence that PSC patients are more susceptible to post-transplant biliary complications with radiological and morphological appearances identical to those of the original disease [110]. The question of whether these cases represent recurrent or acquired sclerosing cholangitis remains unanswered.

8.5 Auto-immune Hepatitis

Recurrence of auto-immune hepatitis (AIH) was first described in a female patient in whom a reduction of the dose of corticosteroids was followed by a rise in serum aspartate aminotransferase (AST), IgG and auto-antibodies. Liver histology showed typical features of CH with a dense mononuclear cell infiltrate in enlarged portal tracts, piecemeal necrosis and some parenchymal bridging collapse. The clinical, serological and histological features resolved as the dose of corticosteroids was increased [80]. In an analysis of 43 patients with AIH, histological evidence of disease recurrence was found in 11 female patients at a median of 18 months after transplantation; 9 of these recurrences received HLA DR3 mismatched grafts [133].

8.6 Recurrence of Primary Liver Tumour

Tumour recurrence has been very high in all early series, and retrospective studies have allowed the identification of factors that were associated with the best results, in particular with the lowest rate of tumour recurrence. Thus, current indications are largely based on past experience which is reviewed briefly here.

8.6.1 Hepatocellular Carcinoma

Early recurrence has occurred in most hepatocellular carcinomas (HCC) larger than 8 cm, in those with multiple nodules affecting both lobes (Stage IVa) and in nearly all cases with lymph-node metastases, gross vascular invasion or extra-capsular growth diagnosed at the time of surgery [100]. These categories, which include the vast majority of patients who present with symptoms related to their tumour, now preclude transplantation, as palliation is better achieved by limited resection. After transplantation, however, tumour growth appears enhanced by the suppression of host immunity [134]. In contrast, the prognosis is excellent with small tumours incidentally found in explanted livers. The recurrence rate has also been low in cirrhotic patients with tumours of less than 4 cm, single or multiple, but unilobar, and who have been diagnosed as part of prospective screening or thorough radiological work-up prior to transplantation [72]. Interestingly, in a retrospective study comparing morphology and imaging, although radiology had been inaccurate in diagnosing small lesions and satellite nodules, it had correctly identified all the lesions that eventually recurred [101].

8.6.2 Fibrolamellar Hepatocellular Carcinoma

The recurrence rate of fibrolamellar hepatocellular carcinoma (FL-HCC) has been unexpectedly high for a tumour, reputedly slow-growing and of high resectability. The absence of symptoms in these non-cirrhotic patients often allows the tumour to grow to considerable size and most patients submitted to OLT have particularly advanced lesions. In addition, histology of the resected tumour often reveals foci that differ from the classical appearances by being occupied by smaller cells with a more basophilic cytoplasm and a scanty fibrous stroma. Whether these foci reflect a more aggressive tumour from the start or a natural transformation in lesions that have been allowed to evolve for prolonged period of time remains to be determined. Despite the high recurrence rate of 40%–83% reported by individual centres, overall survival figures compare favourably with those of non-FL-HCC [73].

8.6.3 Cholangiocarcinoma

The recurrence rate of cholangiocarcinoma (CCA) has been generally higher than that of HCC [85, 91]. Both peripheral and central tumours show a propensity to invade lymphatics at an early stage. Microscopic deposits in perineurial lymphatics is often demonstrable close to the hilar resection of liver-bearing small tumours. The involvement of lymph nodes has been invariably followed by tumour recurrence with such a poor survival rate that a frozen section should be performed and a back-up case should be available at the time of surgery.

8.6.4 Malignant Epithelioid Haemangioendothelioma

Malignant epithelioid haemangioendothelioma (EHAE), mainly observed in soft tissue and lung, may rarely present as a liver tumour with or without evidence of extrahepatic growth. The clinical course is variable, but multiple tumour deposits in both lobes may lead to liver failure, in which case no therapeutic measure has proved effective, short of liver replacement [26]. OLT has been successful, even in patients with evidence of metastasis at the time of surgery [67]. Of four patients transplanted for EHAE in this centre with long-term follow-up, two with discrete tumour nodules affecting both liver lobes are alive and well at 3.5 years and 8.5 years after transplantation. The other two died with tumour recurrence at 9 months and 24 months; both explanted livers showed massive infiltration of both lobes and, in one, the tumour had extensively invaded the porta hepatis. In addition, histology in these two cases revealed atypical tumour foci, namely dilated vascular spaces lined by pleomorphic tumour cells which were reminiscent of an angiosarcoma. Such findings may be associated with a more aggressive tumour behaviour.

8.7 Alcoholic Steato-hepatitis

Figures of 12%–95% have been produced from individual centres regarding the proportion of alcoholic recipients who have resumed alcohol use after OLT. A total of 154 graft biopsy specimens of 23 such patients after transplantation [3] has revealed a spectrum of liver damage, ranging from steatosis, central sclerosis and extensive pericellular fibrosis to cirrhosis. The latter was observed in nine biopsy specimens of four patients obtained from 6 months to 2 years post-transplant. We have observed rapid development of cirrhosis within 2 years of surgery in two patients. However, with the limited number of sequential biopsies available, it is difficult to determine to the respective role of non-alcohol-related graft injury and host-susceptibility factors in this unusually rapid progression.

8.8 Budd-Chiari Syndrome

Recurrence of thromboembolic phenomena have been well recognised and have contributed significantly to the post-operative morbidity and mortality in early series [47]. Vascular thrombosis can affect not only the hepatic veins, but also the portal veins and hepatic artery. Due to the efficacy of other surgical procedures, Budd-Chiari syndrome has now become a rare indication for transplantation that is reserved for those cases with extensive liver-cell loss or fibrosis.

8.9 Metabolic Disorders

Whether metabolic diseases recur after liver transplantation depends on the site of the metabolic defect. When the defect occurs primarily within the hepatocyte, such as in Wilson's disease α1-antitrypsin deficiency, replacement of the liver will result in correction of the disease. Long-term follow-up will be needed to determine whether iron will slowly re-accumulate in recipients with genetic haemochromatosis. Conversely, when the metabolic defect arises outside the liver, the disease state may persist after liver replacement.

Recurrence of the original disease may not necessarily preclude liver replacement as a therapeutic option. If disease recurrence is slow or can be prevented, then grafting remains an appropriate therapy. In addition, it is important to note that even with lysosomal disorders, such Gaucher's disease, Wolman disease and Niemann-Pick disease type A and C, in which enzymatic defects primarily affect the macrophage system and precursors, replacement of failing livers without bone-marrow transplantation has been associated with significant and unpredicted beneficial effects on extrahepatic sites. This is now attributed to the persistence of migratory donor cells thatmay act as enzyme carriers [116].

9 Late Changes of Uncertain Significance

With improvement in patients' survival, an increasing number of allograft biopsies have become available 5 years after surgery. These may present more than one insult, making interpretation difficult [89] or reveal changes of uncertain significance [111].

9.1 Structural Anomaly

Hepatic structural anomalies are commonly observed and include hepatocytic plate disarray, with focal atrophy or thickening, patchy sinusoidal dilatation and hepatocyte anisonucleosis. Rarely, the changes are more marked with parenchymal areas seemingly compressed by somewhat nodular hyperplastic foci, suggesting a diagnosis of nodular regenerative hyperplasia (NRH). Such a diagnosis is difficult on needle biopsy, but best demonstrated on a reticulin preparation (Fig. 13a–c). In a few cases, the full picture of NRH has been shown on subsequently removed grafts (Fig. 13d).

Although mild changes are common and of uncertain significance, there is good evidence that progression to NRH is related to azathioprine (AZA) toxicity. Similar changes have been described after a renal-transplant recipient is maintained on AZA. The changes are more frequent in late biopsies from patients maintained on AZA [111] than in those not receiving AZA [89], and regression of the changes has been observed following AZA withdrawal [40]. Additional factors may be an obliterative portal and/or hepatic venopathy; one third of the patients in our study had potential clotting abnormalities after having received their liver graft for the Budd-Chiari syndrome or were subsequently shown to have thrombotic occlusions of small vascular branches in the graft [111].

9.2 Chronic Hepatitis

Portal lymphocytic infiltrates are common in biopsy specimens taken more than 1 year after transplantation. These were earlier referred to as "late changes" to acknowledge their uncertain aetiology [129]. In some cases, there is sufficient periportal cell spillover, interface and lobular necro-inflammatory changes to diagnose CH, the activity of which may range from minimal to severe and includes a varying degree of lobular cell loss up to parenchymal bridging collapse (Fig. 14a, b). A number of these cases have been subsequently diagnosed as recurrent HCV or HBC infection or AIH, but many patients remain with an undetermined cause for their CH features. A higher proportion of patients had been transplanted for PBC, reflecting either the prevalence of this indication or a possible disease recurrence with a histology lacking the pathognomonic features. In a recent study, fulminant hepatitis of undetermined cause in the native liver was suggested to be a risk factor for post-transplant CH [75]. An unrecognised form of cellular rejection might

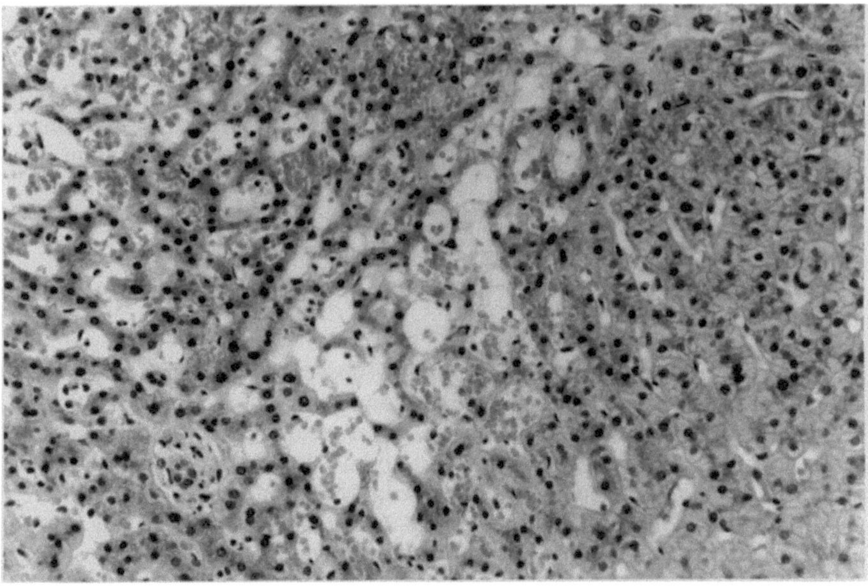

a

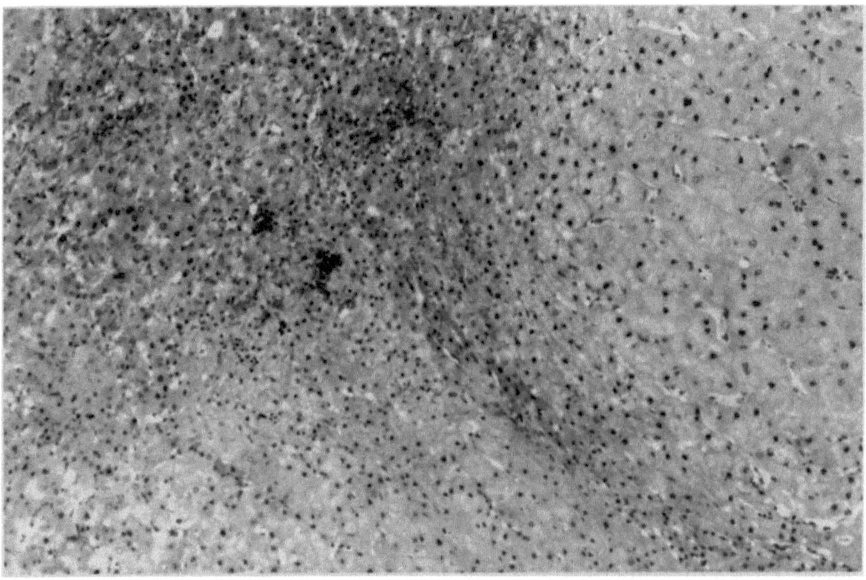

b

Fig. 13 a, b. Structural anomalies presumably related to azathioprine. **a** Sinusoidal dilatation and congestion with concomitant thinning of the intervening hepatocytic plates at 9 months after OLT. **b** Nodular parenchymal hyperplasia which alternates with areas of seemingly compressed parenchyma 2.5 years after OLT. **c, d** see p. 97

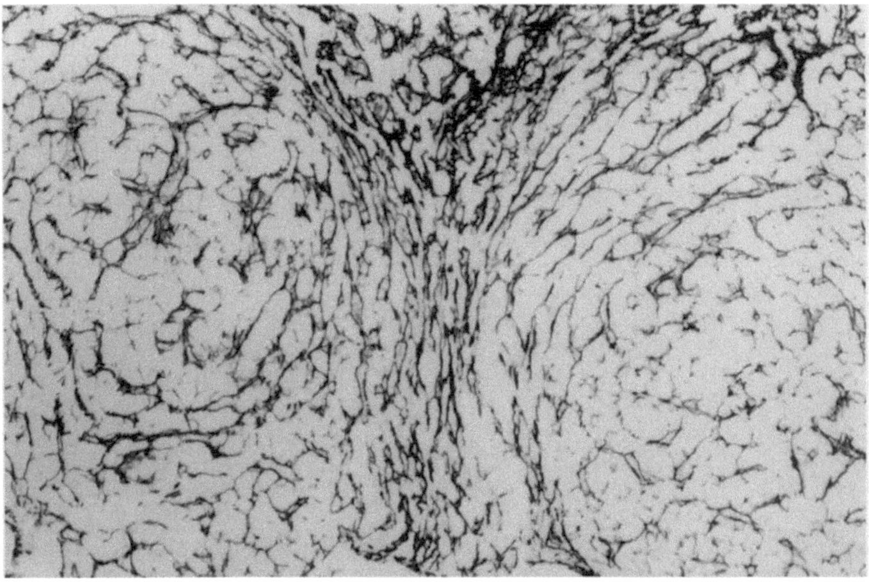

c

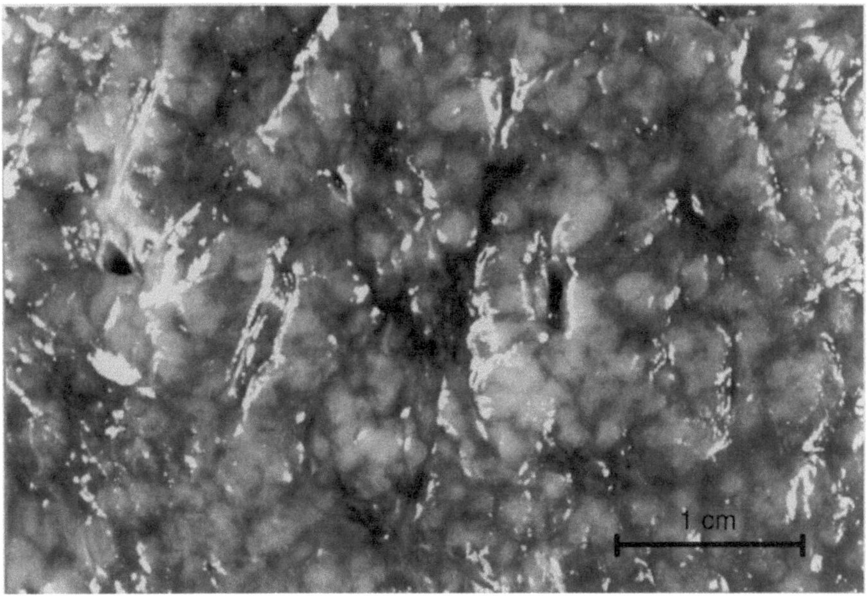

d

Fig. 13. c Similar field as in (*b*), stained for reticulin. **d** Cut section of a liver graft removed 6 years after OLT. There are evenly distributed parenchymal micronodules, which show a tendency to bulge, whereas fibrosis is inconspicuous, a pattern characteristic of nodular regenerative hyperplasia

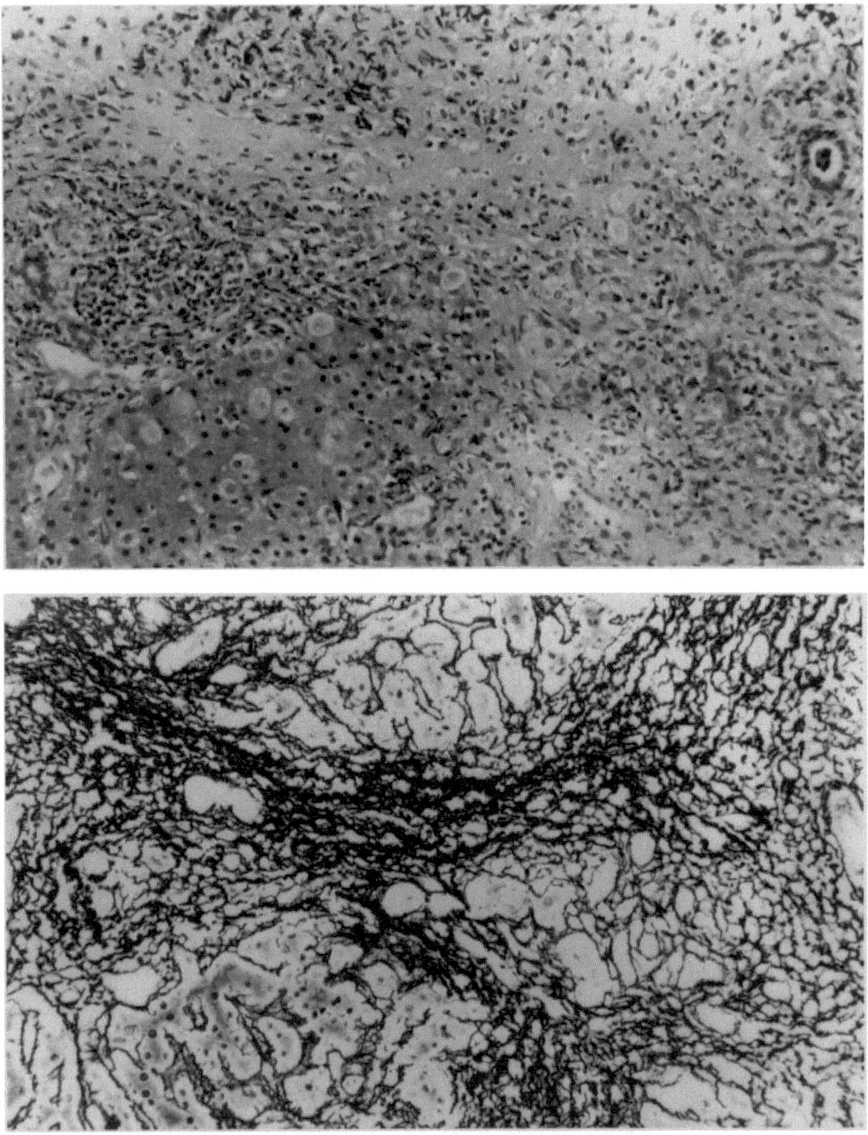

a

b

Fig. 14 a, b. Chronic hepatitis of undetermined aetiology. Parenchymal collapse bridging portal tracts together and to a hepatic venular region with prominent interface hepatitis: **a** H&E; **b** reticulin

be an explanation in other patients who show an immediate response to steroids. Recently, we have found elevated IgG and non-organ-specific autoantibodies in serum of some of these patients transplanted for non-autoimmune disorders. Most of these were children [56].

10 Post-transplant Lymphoproliferative Disorder

Lymphoproliferative disorders (LPD) account for about one fifth of the reported instances of malignancies complicating immunosuppression [91]. Most cases affect B-cells and can be linked with EBV infection [77]. After OLT, primary or reactivated EBV infection of B cells produces a mononucleosis-like disease. In some recipients, impaired T-cell regulation allows EBV-driven B cells to proliferate: (1) into a polyclonal, pseudo-lymphomatous disorder, with tumour-like dissemination in various organs; and (2) more rarely, probably following gene mutation, into a true monoclonal B-cell lymphoma. In situ hybridization and PCR identify the presence of EBV in lymphocytes of the portal infiltrates and of extrahepatic pseudo-lymphomatous or lymphomatous deposits. A reduction in immunosuppression and antiviral drugs may be followed by regression of polyclonal disease, but is less likely to affect monoclonal proliferation; the distinction between the two is desirable. Unfortunately, the monomorphous blastic appearance of some polyclonal proliferations may make distinction from lymphomas difficult. A differential cell count using antibodies to κ and λ light chains is difficult on paraffin sections and, in most instances, confirmation of monoclonality requires gene rearrangement studies.

References

1. Adams DH, Burnett D, Stockley RA, Elias E (1989) Pattern of leucocyte chemotaxis to bile after liver transplantation. Gastroenterology 97: 433–438
2. Angus P, Locarnini S, McCaughan G, Jone R, McMillan J, Bowden D (1995) Hepatitis B virus precore mutant infection in association with severe recurrent disease after liver transplantation. Hepatology 21: 14–18
3. Baddour N, Demetris AJ, Shah G, Tringali R, van Thiel DH (1992) The prevalence, rate of onset and spectrum of histologic liver disease in alcohol abusing liver allograft recipients. Gastroenterology 102: A777
4. Balan V, Batts KP, Porayo MK, Krom RAF, Ludwig J, Wiesner RH (1992) Histologic evidence for recurrence of primary biliary cirrhosis after liver transplantation. Hepatology 18: 1392–1398
5. Benner KG, Lee R G, Keeffe EB, Lopez RR, Sasaki AW, Pinson CW (1992) Fibrosing cytolytic liver failure secondary to recurrent hepatitis B after liver transplantation. Gastroenterology 103: 1307–1312
6. Burke GW, Cirocco R, Viciana A, Ruiz P, Markou M, Allouch M, Cianco G, et al. (1996) Early graft loss secondary to massive hemorrhagic necrosis following orthotopic liver transplantation. Evidence for cytokine-mediated univisceral Shwartzman reaction. Transplantation 61: 1370–1376
7. Cakaloglu Y, Devlin J, O'Grady J, Sutherland S, Portmann BC, Tan K-C, Williams R (1995) Importance of concomitant viral infection during late acute liver allograft rejection. Transplantation 59: 40–45

8. Calne RY, McMaster P, Portmann B, Wall W, Williams R (1977) Observations on preservation, bile drainage and rejection in 64 human orthotopic liver grafts. Ann Surg 186: 282–290
9. Chazouilleres O, Kim M, Combs C, Ferrell L, Bacchetti P, Roberts J, Asher NL, et al. (1994) Quantitation of hepatitis C virus RNA in liver transplant recipients. Gastroenterology 106: 994–999
10. Chisari FV, Filippi P, Buras J, McLachlan A, Popper H, Pinkert CA, Palmiter RD, Brinster RL (1987) Structural and pathological effects of synthesis of hepatitis B virus large envelope polypeptide in transgenic mice. Proc Natl Acad Sci USA 84: 6909–6913
11. Cywes R, Mullen M, Stratis M, Greig P, Levy G, Harvey PRC, Strasberg SM (1993). Prediction of the outcome of transplantation in man by platelet adherence in donor liver allografts. Transplantation 56: 316–323
12. D'Allessandro A, Kalayoglu M, Sollinger H, Hoffman RM, Reed A, Knechtle SJ, Pirsch JD, et al. (1991) The predictive value of donor liver biopsies for the development of primary dysfunction after orthotopic liver transplantation. Transplantation 51: 157–163
13. Datta Gupta S, Hudson M, Burroughs AK, Morris R, Amlot P, Scheuer PJ, Dhillon AP (1995). Grading cellular rejection after orthotopic liver transplantation. Hepatology 21: 46–57
14. Davies HFS, Pollard SG, Calne RY (1989) Soluble HLA antigens in the circulation of liver graft recipients. Transplantation 47: 524–527
15. Davies SE, Portmann BC, O'Grady J G, Aldis PM, Chagar K, Alexander GJM, Williams R (1991) Hepatic histological findings after liver transplantation for chronic hepatitis B virus infection, including a unique pattern of fibrosing cholestatic hepatitis. Hepatology 13: 150–157
16. Davies SE, Lau JYN, O'Grady JG, Portmann BC, Alexander GJM, Williams R (1992) Evidence that hepatitis D virus needs hepatitis B virus to cause hepatocellular damage. Am J Clin Pathol 98: 554–558
17. Demetris AJ, Jaffe R, Sheahan DB, Burnham J, Spero J, Iwatsuki S, Van Thiel, Starzl TE (1986) Recurrent hepatitis B in liver allograft recipients. Differentiation between viral hepatitis B and rejection. Am J Pathol 125: 161–172
18. Demetris AJ, Jaffe R, Starzl TE (1987) A review of adult and pediatric posttransplant liver pathology. In: Somers SC (ed) Pathol Annu 2: 347–386
19. Demetris AJ, Markus B (1989) Immunopathology of liver transplantation. Crit Rev Immunol 9: 67–92
20. Demetris AJ, Fung JJ, Todo S, Banner TB, Zerbe G, Sysyn G, Starzl TE (1990) Pathologic observations in human recipients treated with FK506. Transplant Proc 22 [Suppl 1]: 25–34
21. Demetris AJ, Murase N, Nakamura K, Iwaki Y, Yagishashi A, Valdivia L, Todo S, et al. (1992a) Immunopathology of antibodies as effectors of orthotopic liver allograft rejection. Semin Liver Dis 12: 51–59
22. Demetris AJ, Nakamura K, Yagishashi A, Iwaki Y, Takaya S, Hartman GG, Murase N, et al. (1992b) A clinicopathological study of human allograft recipients harboring preformed IgG lymphocytotoxic antibodies. Hepatology 16: 671–682
23. Demetris AJ, Seaberg EC, Batts KP, Ferrell L Ludwig J, Markin R, Belle S, et al. (1995) Reliability and predictive value of the NIDDK liver transplant database nomenclature and grading system for cellular rejection of liver allografts. Hepatology 14: 751–755
24. Devlin J, Page AC, O'Grady J, Portmann B, Karani J, Williams R (1993) Angiographically determined arteriopathy in liver graft dysfunction and survival. J Hepatol 18: 68–73
25. Devlin J, Smith HM, O'Grady JG, Portmann B, Tan K-C, Williams R (1994) Impact of immunoprophylaxis and patient selection on outcome of transplantation for HBsAg-positive liver recipients. J Hepatol 21: 204–210
26. Dietze O, Davies SE, Williams R, Portmann B (1989) Malignant epithelioid haemangioendothelioma of the liver: a clinicopathological and histochemical study of 12 cases. Histopathology 15: 225–237
27. Dmitrewski J, Hubscher SG, Mayer AD, Neuberger J (1996) Recurrence of primary biliary cirrhosis in the liver allograft: the effect of immunosuppression. J Hepatol 24: 253–257
28. Donaldson P, Underhill J, Doherty D Hayllar K, Calne R, Tan K-C, O'Grady J, Wight D, Portmann B, Williams R (1993) Influence of human leukocyte antigen matching on liver allograft survival and rejection: "the dualistic effect". Hepatology 17: 1008–1015

29. Doussett B, Hübscher S, Padbury R, Gunson BK, Buckels JA, Mayer AD, Elias E, et al. (1993) Acute liver allograft rejection – is treatment always necessary? Transplantation 55: 529–534
30. Dubel L, Farges O, Bismuth H, Sebagh M, Homberg J-C, Johanet C (1995) Kinetics of anti-M2 antibodies after liver transplantation for primary biliary cirrhosis. J Hepatol 23: 674–80
31. Fagan E, Ellis D, Tovey GM, Lloyd G, Smith HM, Portmann B, Tan K-C, et al. (1992) Toga virus-like particles in acute liver failure attributed to sporadic, non-A non-B and recurrence after liver transplantation. J Med Virol 38: 71–77
32. Féray C, Zignego AL, Samuel D,Bismuth A, Reynès M, Tiollais P, Bismuth H, et al. (1990) Persistent hepatitis virus infection of mononuclear cells without concomitant liver infection: the liver transplantation model. Transplantation 49: 1155–1158
33. Féray C, Gigou M, Samuel D, Paradis V, Wilber J, David MF, Urdea M, et al. (1994) The course of hepatitis C virus infection after liver transplantation. Hepatology 20: 1137–1143
34. Ferrell L, Bass N, Roberts J, Ascher N (1992a) Lipopeliosis: fat induced sinusoidal dilatation in transplanted liver mimicking peliosis hepatis. J Clin Pathol 45: 1109–1110
35. Ferrell L, Wright TL, Roberts J, Ascher N, Lake J (1992b) Hepatitis C viral infection in liver transplant recipients. Hepatology 16: 865–876
36. Foster PF, Sankary HN, Hart M, Williams JW, Bhattacharyya A, Coleman J, Ashamann M (1991) Blood and graft eosinophilia as predictors of rejection in human liver transplantation. Transplantation 51: 873–876
37. Freese DK, Snover DC, Sharp HL, Gross CR, Savick SK, Payne WD (1991) Chronic rejection after liver transplantation: a study of clinical, histological and immunological features. Hepatology 13: 882–891
38. Fried MW, Khudyakov YE, Smallwood GA, Cong M-E, Nichols B, Diaz E, Siefert P, et al. (1997) Hepatitis G virus co-infection in liver transplantation recipients with chronic hepatitis C and nonviral chronic liver disease. Hepatology 25: 1271–1275
39. Gane E, Portmann B, Saxena R, Wong P, Ramage J, Williams R. (1994) Nodular regenerative hyperplasia of the liver graft after liver transplantation. Hepatology 21: 88–94
40. Gane E, Portmann B, Naoumov NV, Smith H, Underhill JA, Donaldson PT, Maertens G, Williams R (1996a) Long-term outcome of hepatitis C infection after liver transplantation. N Engl J Med 334: 815–820
41. Gane EJ, Naoumov NV, Qian K-P, Mondelli M, Maertens G, Portmann BC, Lau JYN, Williams R (1996b) A longitudinal analysis of hepatitis C virus replication following liver transplantation. Gastroenterology 110: 167–177
42. Gimson AES, Karani J, Heaton ND (1995) Major biliary tract and vascular complications. In: Williams R, Portmann B, Tan K-C (eds) The practice of liver transplantation. Churchill-Livingstone, Edinburgh, pp 199–209
43. Gouw ASH, Houthoff HJ, Huitema S, Beelen JM, Gips CH, Poppema S (1987) Expression of major histocompatibility complex antigens and replacement of donor cells by recipient ones in human liver grafts. Transplantation 43: 291–296
44. Gouw ASH, Haagsma EB, Manns M, Klompmaker IJ, Slooff MJH, Gerber M (1994) Is there recurrence of primary biliary cirrhosis after transplantation? A clinicopathologic study in long-term survivors. J Hepatol 20: 500–507
45. Greenson JK, Svobodanewman SM, Merion RM, Frank TS (1996) Histologic progression of recurrent hepatitis C in liver transplant allografts. Am J Surg Pathol 20: 731–738
46. Gugenheim J, Samuel D, Reynès M, Bismuth H (1990) Liver transplantation across ABO blood group barriers. Lancet 336: 519–523
47. Halff G, Todo S, Tzakis AG, Gordon RD, Starzl TE (1990) Liver transplantation for Budd-Chiari syndrome. Ann Surg 211: 43–49
48. Harrison RF, Davies MH, Neuberger JM, Hübscher SG (1994) Fibrous and obliterative cholangitis in liver allografts: evidence of recurrent primary sclerosing cholangitis? Hepatology 20: 356–361
49. Henderson JM, Mackay GJ, Lumsden AB, Atta HM, Brouillard R, Kutner MH (1992) The effect of liver denervation on hepatic haemodynamics during hypovolaemic shock in swine. Hepatology 15: 130–133

50. Hübscher SG, Adams DH, Buckels JAC, McMaster P, Neuberger J, Elias E (1989) Massive haemorrhagic necrosis of the liver after liver transplantation. J Clin Pathol 42: 60–370
51. Hübscher SG, Buckels JAC, Elias E, McMaster P, Neuberger J (1991) Vanishing bile-duct syndrome following liver transplantation-Is it reversible? Transplantation 51: 1004–1010
52. International Panel (1997) Banff schema for grading liver allograft rejection: an international consensus document. Hepatology 25: 658–663
53. International Working Party (1995) Terminology for hepatic allograft rejection. Hepatology 22: 648–654
54. Kakizoe S, Yanaga K, Starzl TE, Demetris AJ (1990) Evaluation of protocol before transplantation and after reperfusion biopsies from human orthotopic liver allografts: considerations of preservation and early immunological injury. Hepatology 11: 932–941
55. Kemnitz J, Gubernatis G, Bunzendahl H, Ringe B, Pichlmayr RAG (1989) Criteria for the histopathological classification of liver allograft rejection and their clinical relevance. Transplant Proc 21: 2208–2210
56. Kerkar N, Hadzic N, Davies ET, Portmann B, Donaldson PT, Rela M, Heaton ND, Vergani D, Mieli-Vergani G (1998) De novo "autoimmune" hepatitis after transplantation Lancet 351: 409–413
57. Klintmalm G, Nery J, Husberg B, Gonwa TA, Tillery GW (1989) Rejection in liver transplantation. Hepatology 10: 978–985
58. Koskinas J, Portmann B, Lombard M, Smith T, Williams R (1992) Persistent iron overload 4 years after inadvertent transplantation of a heamochromatosis liver in a patient with primary biliary cirrhosis. J Hepatol 16: 351–354
59. Langnas AN, Marujo W, Stratta RW, Wood RP, Shaw BW (1991) Vascular complications after liver transplantation. Am J Surg 23: 76–79
60. Lautenschlager I, Höckerstedt K, Häyry P (1991) Fine-needle aspiration biopsy in the monitoring of liver allografts. Transpl Int 4: 54–61
61. Leon MP, Bassendine MF, Gibbs P, Burt AD, Thick M, Kirby JA (1996) Hepatic allograft rejection: regulation of the immunogenicity of human intrahepatic biliary epithelial cells. Liver Transpl Surg 1: 37–45
62. Li S, Stratta RJ, Langnas AN, Wood RP, Marujo W, Shaw BW (1992) Diffuse biliary tract injury after orthotopic liver transplantation. Am J Surg 164: 536–540
63. Ludwig J, Wiesner R, Batts K, Perkins JD, Krom RAF (1987) The acute vanishing bile duct syndrome (acute irreversible rejection) after orthotopic liver transplantation. Hepatology 7: 476–483
64. Ludwig J, Batts KP, MacCarty RL (1992) Ischaemic cholangitis in hepatic allograft. Mayo Clin Proc 67: 519–526
65. Ludwig J, Batts KP (1994) Transplantation pathology. In: MacSween RNM, Anthony PP, Scheuer PJ, Burt AD, Portmann BC (eds) Pathology of the liver, 3rd edn. Churchill Livingstone, Edinburgh, pp 765–796
66. MacDonald GA, Greenson JK, Del Buono EA, Grady WM, Merion RM, Frank TS, Lucey MR, Appleman HD (1997) Mini-microabscess syndrome in liver transplant recipients. Hepatology 26: 192–197
67. Marino IR, Todo S, Tzakis A, Klintmalm G, Kelleher M, Iwatsuki S, Starzl TE, Esquivel CO (1988) Treatment of hepatic epithelioid hemangioendothelioma with liver transplantation. Cancer 62: 2079–2084
68. Marsh JW, Iwatsuki S, Makowka L, Esquivel CO, Gordon RD, Todo S, Tzakis A, et al. (1988) Orthotopic transplantation for primary sclerosing cholangitis. Ann Surg 207: 21–25
69. Matsumoto Y, McCaughan GW, Painter DM, Bishop A (1993) Evidence that portal tract microvascular destruction precedes bile duct loss in human liver allograft rejection. Transplantation 56: 69–75
70. Mazariegos GV, Reyes J, Marino IR, Demetris AJ, Flynn B, Irish W, McMichael J, et al. (1997) Weaning immunosuppression in liver transplant recipients. Transplantation 63: 243–249
71. McCaughan GW, Davies JS, Waugh JA, Bishop GA, Hall BM, Gallagher ND, Thompson JF, et al. (1990) A quantitative analysis of T lymphocyte populations in human liver allografts undergoing rejection: the use of monoclonal antibodies and double immunolabeling. Hepatology 12: 1305–1313

72. McPeake JR, O'Grady JG, Zaman S, Portmann B, Wight DGD, Tan K-C, Calne RY, Williams R. (1993) Liver transplantation for primary hepatocellular carcinoma: tumour size and number determine outcome. J Hepatol 18: 226–234
73. McPeake JR, Portmann B (1995) Hepatic malignancy, Budd-Chiai syndrome and space occupying conditions. In: Williams R, Portmann B, Tan K-C (eds) The practice of liver transplantation. Churchill-Livingstone, Edinburgh, pp 57–71
74. Michaels MG, Green M, Wald ER, Starzl TE (1992) Adenovirus infection in pediatric liver transplant recipients. J Infect Dis 165: 170–174
75. Mohamed R, Hubscher SG, Mirza DF, Gunson BK, Mutimer DJ (1997) Posttransplantation chronic hepatitis in fulminant hepatic failure. Hepatology 25: 1003–1007
76. Mondelli M, Mieli-Vergani G, Alberti A Vergani D, Portmann B, Eddleston ALWF, Williams R (1982) 1982 Specificity of T lymphocyte cytotoxicity to autologous hepatocytes in chronic hepatitis B virus infection: evidence that T cells are directed against HBV core antigen expressed on hepatocytes. J Immunol 129: 2773–2778
77. Nalesnik MA, Jaffe R, Starzl TE, Demetris AJ, Porter K, Burnham JA, et al. (1988) The pathology of posttransplant lymphoproliferative disorders occurring in the setting of cyclosporine A-prednisone immunosuppression. Am J Pathol 133: 173–192
78. Naoumov NV, Alexander GJM, O'Grady JG, Aldis P, Portmann BC, William R (1988) Rapid diagnosis of cytomegalovirus infection by in-situ hybridisation in liver grafts. Lancet 1: 1361–1364
79. Neuberger JM, Portmann B, MacDougall B, Calne RY, Williams R (1982) Recurrence of primary biliary cirrhosis after liver transplantation. N Engl J Med 306: 1–4
80. Neuberger J, Portmann B, Calne R, Williams R (1984) Recurrence of autoimmune chronic active hepatitis following orthotopic liver transplantation. Transplantation 37: 363–365
81. Neuberger J (1995) Liver allograft rejection - current concepts on diagnosis and treatment. J Hepatol 23[Suppl 1]: 54–61
82. Ng IOL, Burroughs AK, Rolles K, Belli LS, Scheuer PJ (1991) Hepatocellular ballooning after liver transplantation: a light and electronmicroscopic study with clinicopathological correlation. Histopathology 18: 323–330
83. Noack KB, Wiesner RH, Batts K, van Hoek B, Ludwig J (1991) Severe ductopenic rejection with features of vanishing bile duct syndrome: clinical, biochemical, and histologic evidence for spontaneous resolution. Transplant Proc 23: 1448–1451
84. O'Grady JG, Alexander GJM, Sutherland S Donaldson PT, Harvey F, Portmann B, Calne RY, Williams R (1988a) Cytomegalovirus infection and donor/recipient HLA antigens: interdependent co-factors in pathogenesis of vanishing bile duct syndrome after transplantation. Lancet 2: 302–305
85. O'Grady JG, Polson RJ, Rolles K, Calne RY, Williams R (1988b) Liver transplantation for malignant disease. Results in 93 consecutive patients. Ann Surg 207: 373–379
86. O'Grady JG, Smith HM, Davies SE, Daniels HM, Donaldson PT, Tan K-C, Portmann B, et al. (1992) Hepatitis B reinfection after orthotopic liver transplantation. Serological and clinical implication. J Hepatol 14: 104–111
87. Oguma S, Belle S, Starzl TE, Demetris AJ (1989) A histometric analysis of chronically rejected human liver allografts: insights into the mechanism of bile duct loss: direct immunologic and ischaemic factors. Hepatology 9: 204–209
88. Ottobrelli A, Marzano A, Smedile A, Recchia S, Salizzoni M, Cornu C, Lamy ME, et al. (1991) Patterns of hepatitis Delta virus reinfection and disease in liver transplantation. Gastroenterology 101: 1649–1655
89. Pappo O, Ramos H, Starzl T, Fung J, Demetris A (1995) Structural integrity and identification of causes of liver allograft dysfunction occurring more than 5 years after transplantation. Am J Surg Pathol 19: 192–206
90. Paya C, Holley K, Wiesner RH, Balasubramaniam K, Smith TF, Espy MJ, Ludwig J, et al. (1990) Early diagnosis of CMV hepatitis in liver transplant recipients: role of immunostaining, DNA hybridisation and culture of liver tissue. Hepatology 12: 119–126
91. Penn I (1990) Cancers complicating organ transplantation. N Engl J Med 323: 1767–1769
92. Phillips MJ, Cameron R, Flowers MA, Blendis LM, Greig PD, Wanless I, Sherman M, et al. (1992) Post-transplant recurrent hepatitis B viral disease; viral-burden, steatoviral, and fibroviral hepatitis B. Am J Pathol 40: 1295–1308

93. Polson R, Portmann B, Neuberger JM, Calne RY, Williams R (1989) Evidence for disease recurrence after liver transplantation for primary biliary cirrhosis. Gastroenterology 97: 715–725
94. Portmann B, Gane E (1995) Use of liver biopsy. In: Williams R, Portmann B, Tan K-C (eds) The practice of liver transplantation. Churchill-Livingstone, Edinburgh, pp 225–241
95. Portmann B, Neuberger JM (1995) Disease recurrence (viral and other disorders). In: Williams R, Portmann B, Tan K-C (eds) The practice of liver transplantation. Churchill-Livingstone, Edinburgh, pp 245–257
96. Portmann B, Neuberger JM, Williams R (1983) Intrahepatic bile duct lesions. In: Calne RY (ed) Liver transplantation. Grune & Stratton, London, pp 279–287
97. Portmann B, Wight DGD (1987) Pathology of liver transplantation (excluding rejection). In: Calne RY (ed) Liver transplantation, 2nd edn. Grune and Stratton, London, pp 437–470
98. Powell LW (1992) Does transplantation of the liver cure genetic haemochromatosis? J Hepatol 16: 259–261
99. Read AE, Donegan E, Lake J, Ferrell L, Galbraith C, Kuramoto IK, Zeldis JB, et al. (1991) Hepatitis C in patients undergoing liver transplantation. Ann Intern Med 114: 282–284
100. Ringe B, Pichlmayr R, Wittekind R, Tusch G (1991) Surgical treatment of hepatocellular carcinoma: experience with liver resection and transplantation in 198 patients. World J Surg 15: 270–285
101. Rizzi PM, Kane P, Ryder SD, Ramage JK, Gane E, Tan K-C, Portmann B, Karani J, Williams R (1994) Accuracy of radiology in detection of HCC prior to transplantation: comparison with explanted liver. Gastroenterology 107: 1425–1429
102. Sallie R, Silva A E, Purdy M, Smith H, McCaustland K, Tibbs C, Portmann B, et al. (1994) Hepatitis C and E in non-A non-B fulminant hepatic failure: a polymerase chain reaction and serological study. J Hepatol 20: 580–588
103. Samuel D, Bismuth A, Mathieu D, Arulnaden J-L, Reynès M, Benhamou J-P, Bréchot C, Bismuth H (1991) Passive immunoprophylaxis after liver transplantation in HBsAg-positive patients. Lancet 1: 813–815
104. Samuel D, Muller R, Alexander G, Fassati L, Ducot B, Benhamou J-P, Bismuth A and European Concerted Action on Viral Hepatitis (EUROHEP) (1993) Liver transplantation in European patients with the hepatitis B surface antigen. N Engl J Med 329: 1842–1847
105. Sanchez-Urdazpal L, Gores GJ, Ward EM, Maus TP, Buckel EG, Steers JL, Wiesner RH, Krom RAF (1993) Diagnostic features and clinical outcome of ischemic-type biliary complications after liver transplantation. Hepatology 17: 605–609
106. Saxena R, Tovey DG, Dhawan A, Ellis DDS, Portmann BC (1996) Acute liver failure due to adenovirus hepatitis in a paediatric liver transplant. Int J Surg Pathol 3: 189–193
107. Schluger LK, Sheiner PA, Thung SN, Lau JYN, Min A, Wolf DC, Fieal I, et al. (1996) Severe cholestatic hepatitis C following orthotopic liver transplantation. Hepatology 23: 971–976
108. Sebagh M, Farges O, Kalil A, Samuel D, Bismuth H, Reynes M (1995) Sclerosing cholangitis following human orthotopic liver transplantation. Am J Surg Pathol 19: 81–90
109. Shah G, Demetris AJ, Gavaler JS, Lewis JH, Todo S, Starzl TE, van Thiel DH (1992) Incidence, prevalence, and clinical course of hepatitis C following liver transplantation. Gastroenterology 103: 323–329
110. Sheng R, Campbell WL, Zajko AB, Baron RL (1996) Cholangiographic features of biliary strictures after liver transplantation for primary sclerosing cholangitis: evidence for recurrent disease. AJR Am J Roengenol 166: 1109–1113
111. Slapak GI, Saxena R, Portmann B, Gane E, Calne R, Williams R (1997) Graft and systemic disease in long-term survivors of liver transplantation. Hepatology 25: 195–202
112. Snover DC, Sibley RK, Freese DK, Sharp HL, Bloomer JR, Najarian JS, Ascher N (1984) Orthotopic liver transplantation: a pathological study of 63 serial liver biopsies from 17 patients with special reference to the diagnostic features and natural history of rejection. Hepatology 4: 1212–1222
113. Snover DC, Freese DK, Sharp HL, Bloomer JR, Najarian JS, Ascher NL (1987) Liver allograft rejection. An analysis of the use of biopsy in determining the outcome of rejection. Am J Surg Pathol 11: 1–10

114. Starzl TE, Putnam CW, Hansbrough JF, Porter KA, Reid HAS (1977) Biliary complications after liver transplantation: with special reference to the biliary cast syndrome and technique of secondary duct repair. Surgery 81: 212–221
115. Starzl TE, Demetris AJ, Todo S, Kang Y, Tzakis A, Dusquesnoy R, Makowka L, et al. (1989) Evidence for hyperacute rejection of human liver grafts: the case of the canary kidneys. Clin Transpl 3: 37–48
116. Starzl TE, Demetris AJ, Trucco M, Murase N, Ricordi C, Ildstad S, Ramos H, et al. (1993) Cell migration and chimerism after whole-organ transplantation. The basis of graft acceptance 17: 1127–1152
117. Steinhoff G, Wonigeit K, Pichlmayr R (1988) Analysis of sequential changes in major histocompatibility complex expression in human liver grafts after transplantation. Transplantation 45: 394–401
118. Stratta RJ, Wood RP, Langnas AN, Hollins RR, Bruder KJ, Donovan JP, Burnett DA, et al. (1989) Diagnosis and treatment of biliary tract complications after orthotopic liver transplantation. Surgery 106: 675–684
119. Telenti A, Smith TF, Ludwig J, Ludwig J, Keating MR, Krom RAF, Wiesner RH (1991) Epstein-Barr virus and persistent graft dysfunction after liver transplantation. Hepatology 14: 282–286
120. Terrault NA, Wright TL (1997) Leading article. Hepatitis B virus infection and liver transplantation. Gut 40: 568–571
121. Thung SN, Shim K-S, Shieh C, Schwartz M, Theise N, Borcich A, Katz E, et al. (1993) Hepatitis C in liver allografts. Arch Pathol Lab Med 117: 145–149
122. Todo S, Demetris A, Makowka L, et al. (1989) Primary nonfunction of hepatic allografts with pre-existing fatty infiltration. Transplantation 47: 903–905
123. Todo S, Demetris AJ, van Thiel D, Teperman L, Fung J, Starzl TE (1991) Orthotopic liver transplantation for patients with hepatitis B virus-related liver disease. Hepatology 13: 619–626
124. Torisu M, Yokoyama T, Amemiya H, Kohler PF, Schroter G, Martineau G, Penn I, et al. (1971) Immunosuppression, liver injury, and hepatitis in renal, hepatic, and cardiac homograft recipients: with particular reference to the Australia antigen. Ann Surg 174: 620–639
125. Turlin B, Hayllar KM, Slapak GI, Heaton N, Williams R, Portmann B (1995) Centrilobular necrosis after orthotopic liver transplantation: a longitudinal clinicopathologic study in 71 patients. Liver Transpl Surg 1: 285–289
126. Van de Water J, Gerson LB, Ferrell LD, Lake JR, Coppel RL, Batts KP, Wiesner RH, Gershwin ME (1996) Immunohistochemical evidence of disease recurrence after liver transplantation for primary biliary cirrhosis. Hepatology 24: 1079–1084
127. Van Hoek B, Wiesner R, Krom R, Ludwig J, Moore SB (1992) Severe ductopenic rejection following liver transplantation: incidence, time of onset, risk factors, treatment and outcome. Semin Liver Dis 12: 41–50
128. Van Thiel DH, Schade RR, Gavaler JS, Shaw BW, Iwatsuki S, Starzl TE (1984) Medical aspects of liver transplantation. Hepatology 4: 79S-83S
129. Wight DGD, Portmann B (1987) Pathology of liver transplantation. In: Sir Roy Calne (ed) Liver transplantation, 2nd edn. Grune & Stratton, London, pp 385–435
130. Williams J, Vera S, Peters T, Van Voorst S, Britt LG, Dean PJ Haggitt R, et al. (1986) Cholestatic jaundice after hepatic transplantation. Am J Surg 151: 65–69
131. Wong P, Devlin J, Gane E, Ramage J, Portmann B, Williams R (1993) FK506 rescue therapy for intractable rejection. J Hepatol 17: 284–287
132. Wright TL, Donegan E, Hsu HH, Ferrell L, Lake JR, Kim M, Combs C, et al. (1992a) Recurrent and acquired hepatitis C viral infection in liver transplant recipients. Gastroenterology 103: 317–322
133. Wright HL, Bou-Abboud C, Hassanein T (1992b) Disease recurrence following liver transplantation for autoimmune chronic active liver disease. Transplantation 53: 136–139
134. Yokoyama I, Carr B, Saitsu H, Iwatsuki S, Starzl TE (1991) Accelerated growth rates of recurrent hepatocellular carcinoma after liver transplantation. Cancer 68: 2095–2100

Bone-Marrow and Blood Stem-Cell Transplantation

K. Atkinson

Current Topics in Pathology
Volume 92, C. L. Berry (Ed.)

1 Introduction

The last 10 years have seen a dramatic and dynamic expansion in the clinical discipline of bone-marrow and blood stem-cell transplantation, and an increased understanding of the underlying haematological and immunological science, particularly with regard to cellular and molecular biology. During this time, allogeneic marrow transplantation, using family-member donors has become much safer and, thus, more effective. A well-functioning global volunteer marrow-donor registry network has come on-line and, most dramatically, autologous transplantation is overtaking allogeneic transplantation in numbers. This is due in part to the increased safety of autologous blood stem cell as opposed to autologous marrow transplantation and in part to the demonstration of efficacy of the procedure in patients with lymphoma and breast cancer.

Bone-marrow transplantation involves the administration of high-dose cytotoxic chemotherapy or chemotherapy and total body irradiation in an attempt to destroy the patient's underlying disease. In doing so, the marrow and immune system are ablated and need to be replaced by an infusion of haematopoietic stem cells (HSC). The source of such cells can either be from the patients themselves (autologous transplant), from an identical twin (syngeneic transplant) or from a genetically different but usually human leucocyte antigen (HLA)-matched individual (allogeneic transplant). The different types of transplants are summarized in Table 1, the diseases for which marrow transplantation can be utilized in Table 2 and the key differences between autologous and allogeneic transplants in Table 3.

Autologous transplantation works primarily through the administration of the high-dose chemotherapy or chemoradiation. In contrast, allogeneic transplantation involves, in addition, an immune attack on tumour cells not eradicated by the high-dose chemotherapy that is mediated by donor cells. This is the graft-versus-tumour effect, best known in its graft-versus-leukaemia form.

While bone-marrow HSCs have been the main source of HSCs for many years, peripheral blood HSCs are now the main source for autologous transplantation and are being used increasingly in allogeneic transplantation. Harvesting bone marrow cells requires a general anaesthetic for the donor, during which marrow is physically aspirated through needles inserted percutaneously into the posterior superior iliac crests. It has been known for many years that very low levels of HSCs

Table 1. Types of haematopoietic stem-cell (HSC) transplants

Type of transplant	Donor of HSC	HLA-match between HSC and recipient	Source of HSC
Autologous	Self (patient)	By definition	Bone marrow or blood
Syngeneic	Identical twin	By definition	Bone marrow or blood
Allogeneic	Genetically different (except for MHC genes) person: relatives or unrelated donor	May be 6 of 6 HLA-identical or 5 of 6 HLA-identical	Bone marrow, blood or cord blood

MHC, major histocompatability complex; HLA, human leucocyte antigen.

Table 2. Diseases that can be treated by haematopoietic stem-cell (HSC) transplantation

Disease	Commonest type of transplant used	
	Autologous	Allogeneic
AML	++	++
ALL	+	++
CML	+	+++
CLL	+	++
NHL	+++	+
HD	+++	+
MM	+++	+
MDS	–	++
MF	–	++
SAA	–	+++
Fanconi	–	++
Thalassaemia major	–	+++
SCD	–	+++
SCID/ID	–	++
Metabolic diseases	–	++
Breast cancer	+++	–
Testicular cancer	+++	–
Ovarian cancer	+++	–
Neuroblastoma	+++	–

AML, acute myeloid leukaemia; SAA, severe aplastic anaemia; ALL, acute lymphoblastic leukaemia; SCD, sickle cell disease; CML, chronic myeloid leukaemia; SCID/ID, severe combined immune deficiency; CLL, chronic lymphatic leukaemia; deficiency/other congenital; NHL, non-Hodgkin's lymphoma; HD, Hodgkin's disease; MM, multiple myeloma; MDS, myelodysplasia; MF, myelofibrosis.

Table 3. Features of the different types of transplant

	Autologous	Allogeneic
Stem cells cryopreserved	Yes	No
Transplant - related risk	Low (< 5 %)	Moderate–high (15 – 40 %)
Risk of relapse	Moderate–high	Low–moderate
Chance of cure	Depends on disease	Depends on disease
Graft-versus-tumour effect	No	Yes
Cost	↓	↓

circulate in the blood under normal conditions, but this number can be greatly increased by administration of a haematopoietic growth factor (HGF), such as G-CSF or GM-CSF, by cytotoxic chemotherapy or by a combination of the two. When these cells are present in the blood in large numbers, they can be harvested by apheresis, using a cell-separator machine and requires cannulation of a vein in each arm. This 3-to 4-h collection process, however, is performed on an outpatient basis and requires no general anaesthetic.

To undergo autologous transplantation, the patient's marrow should be in remission or excellent partial remission in order to minimize the risk of tumour cells being collected, cryopreserved and subsequently infused into the recipient after the high-dose chemotherapy. Since current chemotherapy regimes normally last 3–7 days and HSCs cannot be preserved for longer than approximately 48 h at 4°C, autologous bone marrow or blood cells need to be collected and cryopreserved prior to the administration of high-dose chemotherapy. Allogeneic cells, utilizing a normal donor, in contrast, are normally given fresh. Allogeneic donors are most commonly HLA-matched siblings; a given individual has an approximate 30% chance of finding such a donor. Less frequently, the donor may be an HLA-matched or partially matched family member, other than a sibling; the patient will have an approximate 1% chance of finding such a donor. Increasing use is now being made of unrelated donors who are normally matched with the recipient for at least 5 or 6 of the HLA antigens encoded at the HLA-A, HLA-B and HLA-DR loci. Currently, the global donor volunteer tissue-typed registry has approximately 3.0 million such donors and, while the chance of finding a match is high (at least for Caucasians) (70%), such transplants are less frequently performed today than family member transplants because of the high transplant-associated risk involved.

Autologous transplants have a low treatment-related mortality, similar to that of conventional cytotoxic chemotherapy. There is still considerable risk attached to allogeneic transplantation, primarily because of graft-versus-host disease (GVHD) and infection. While autologous transplantation is performed, at least in some centers, primarily on an outpatient basis, this is not yet the case for allogeneic transplantation and the difference in cost is due partly to this. Additionally, immune suppression needs to be administered for 6–12 months to allogeneic transplant recipients to minimize the risk of both graft rejection and GVHD. This does not apply to autologous transplantation.

2 History of Clinical Bone-Marrow Transplantation

In 1957, Thomas et al. [1] reported that large amounts of marrow could be infused safely into patients and described transient marrow engraftment. Mathé and his group [2] in Paris were the most active clinical group during the 1960s and, in 1963, reported the first case of complete engraftment with survival beyond 1 year. This patient with leukaemia developed acute and chronic GVHD and eventually died, free of leukaemia, 20 months later of varicella encephalitis. Dausset [3] described the first HLA-antigen in 1958 (HLA-A2). By 1968, the closely linked loci HLA-A and HLA-B were established and by 1971 the HLA-C locus was identified. These developments in the late 1960s and early 1970s set the stage for a more logical way of choosing a marrow donor than was previously possible.

Storb, Thomas and the Seattle group utilized the canine model to determine the important predictive value of histocompatibility testing on transplant outcome and devised regimens for control of GVHD in dogs given dog leucocyte antigen (DLA)-identical littermate marrow grafts. Based on these studies, the first electively HLA-matched transplants were performed in Seattle for aplastic anaemia [5] and for patients with end stage acute leukaemia [4]. In 1977, the Seattle group reported the results of HLA-identical sibling transplants in 100 patients with end-stage leukaemia [6]. This now classical study illustrated the potential curative effect

Table 4. Historical landmarks in bone-marrow transplantation (BMT) development from 1980 onwards

1. Combination immune suppression (CSP/MTX) to control acute GVHD in HLA-matched sibling recipients
2. Prophylactic ganciclovir to prevent CMV disease
3. Use of autologous blood stem-cell transplants mobilized with haematopoietic growth factors or chemotherapy or both
4. Use of HLA-matched unrelated donors
5. Use of cytokines to accelerate haematopoietic recovery post-transplant
6. Use of cord blood transplants
7. HSC selection
8. Cellular immunotherapy with allogeneic lymphocytes for disease relapse or for some viral infections post-transplant
9. Ex vivo expansion of HSC prior to infusion
10. Gene transduction of HSC prior to infusion
11. Allogeneic blood stem-cell transplants.
12. Transplantation for autoimmune disease

CSP, cyclosporin; MTX, methotrexate; GVHD, graft-versus-host disease; CMV, cytomegalovirus.

of marrow transplantation in acute leukaemia with 13 long-term disease-free survivors. Enthusiasm over this remarkable result, however, was tempered by the actuarial relapse rate of 70% and the high incidence of non-leukaemic deaths. It was reasoned that transplants performed in early remission should fare better, since patients would have a smaller tumour burden and would be in better clinical condition than end-stage patients. The report by the Seattle group in 1979 of transplantation of patients with acute non-lymphoblastic leukaemia in first remission confirmed this [7]. It is generally agreed that the modern era of bone-marrow transplantation began in the 1970s, when current HLA-typing became available and when patients with acute leukaemia were subsequently transplanted in first remission rather than in relapse.

Other landmarks in the development of bone-marrow and blood HSC transplantation over the last two decades are described in Table 4.

3 Biology of Haematopoietic Stem Cell (HSC) Transplantation

3.1 Marrow Ablation: The Need for Marrow Space, Immune Suppression and Malignant Cell Eradication

The elements needed for donor-origin haematopoiesis in allografted patients with leukaemia are shown in Table 5. Space is needed in the recipient's marrow to enable donor-origin stem cells to repopulate the patient's ablated marrow. The recipient's immune system also needs to be ablated to minimize the risk of marrow-graft rejection. This is achieved by the use of the same high-dose chemotherapy or chemoradiation conditioning regimen.

Finally, the underlying disease (for example, malignancy) needs to be eradicated if the patient is to be cured. This is achieved by a combination of the high-dose

Table 5. Elements needed for donor-origin haematopoiesis in allografted patients with leukaemia

1.	Marrow space for donor stem cells (by marrow ablation).
2.	Eradication of recipient immune system (by immune suppression).
3.	Eradication of underlying marrow malignancy (by marrow ablation).

marrow-ablative conditioning regimen together, in the case of allografted patients, with a graft-versus-leukaemia immune attack by donor T cells on residual recipient malignant cells.

3.2 The Haematopoietic Stem Cell

The HSC constitutes approximately 1 in every 2000 bone marrow cells and has a 2000-fold increase in the ability to confer protection from marrow-lethal doses of radiation. The human phenotype is $CD34^{+}$ Thy-1^{lo} ($CDw90^{lo}$), lineage^{-}, rhodamine^{123lo} c-Kit^{+}, HLA-$DR^{+/-}$ $CD38^{+/-}$; (rhodamine^{123} is a mitochondrial dye, the uptake of which correlates with self-renewal capacity).

Characteristics of the human multi-potential HSC are multi-lineage differentiation, self-renewal capacity and the ability to reconstitute a marrow-ablated patient. Lineage negativity includes absence of the items detailed in Table 6. Human HSCs are normally selected because of their characteristic expression of the CD34 cell surface antigen.

3.3 The Bone-Marrow Microenvironment

Developing haematopoietic cells of both myeloid and lymphoid lineages exist, in vivo, in contact with the bone-marrow stroma. The cells and the extracellular matrix (ECM) of this tissue may contribute to the survival and differentiation of HSCs through close-range interactions with the haematopoietic cells. This so called paracrine support is thought to be mediated by a number of HGFs and growth regulatory molecules present on the surface of stromal cells or found associated with the stromal ECM.

Table 6. Lineage negativity

Lineage	Cell-surface antigens
T cell	CD7, 2, 3, 4, 8
B cell	CD10, 19
NK cell	CD56, 16
Myeloid cells	CD33, 15
Erythroid cells	Transferrin (CD71), glycophorin

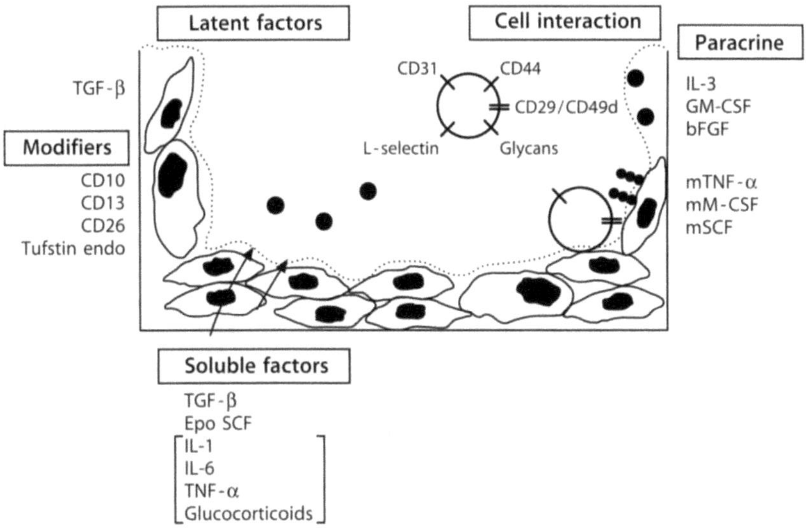

Fig. 1. A schematic diagram of the features of the marrow micro-environment. Adapted with permission from K. Atkinson, editor of Clinical bone-marrow transplantation: a reference textbook (1994) Cambridge University Press, Cambridge

Haematopoietic cells interact with stromal cells and the ECM via an array of adhesion molecules, enabling early haematopoietic cells to localize next to cells and the ECM, which present HGFs (see "*Cell interaction*" box, Fig. 1).

The haematopoietic environment also appears to have an array of mechanisms by which the activity of factors that influence haematopoiesis can be potentiated or attenuated. It is known, for example, that a number of cell-surface proteases are abundant on haematopoietic stroma, including CD10, the common acute lymphoblastic leukaemia antigen. These proteases can act upon factors that influence haematopoiesis either by activating "pro" molecules or by degrading active species (see "*Modifiers*" box, Fig. 1).

In addition to paracrine support, a number of other factors influence the haematopoietic environment (see "*Soluble factors*" box, Fig. 1). Both the haematopoietic progenitors or the marrow stroma can be influenced by serum levels of circulating cytokines, hormones or other factors, for example, erythropoietin, stem-cell factor and glucocorticosteriods. These factors can be generated basally or induced under stress conditions. In this way, marrow output can be modulated by systemic events such as infection or anoxic stress.

3.4 Haematopoietic Growth Factors (HGFs)

Understanding of the molecules controlling haematopoiesis and lymphopoiesis has advanced markedly in the last 15 years. The key control components are the HGFs. These are shown in Table 7. Molecular control of human haematopoiesis by these factors is illustrated in Fig. 2.

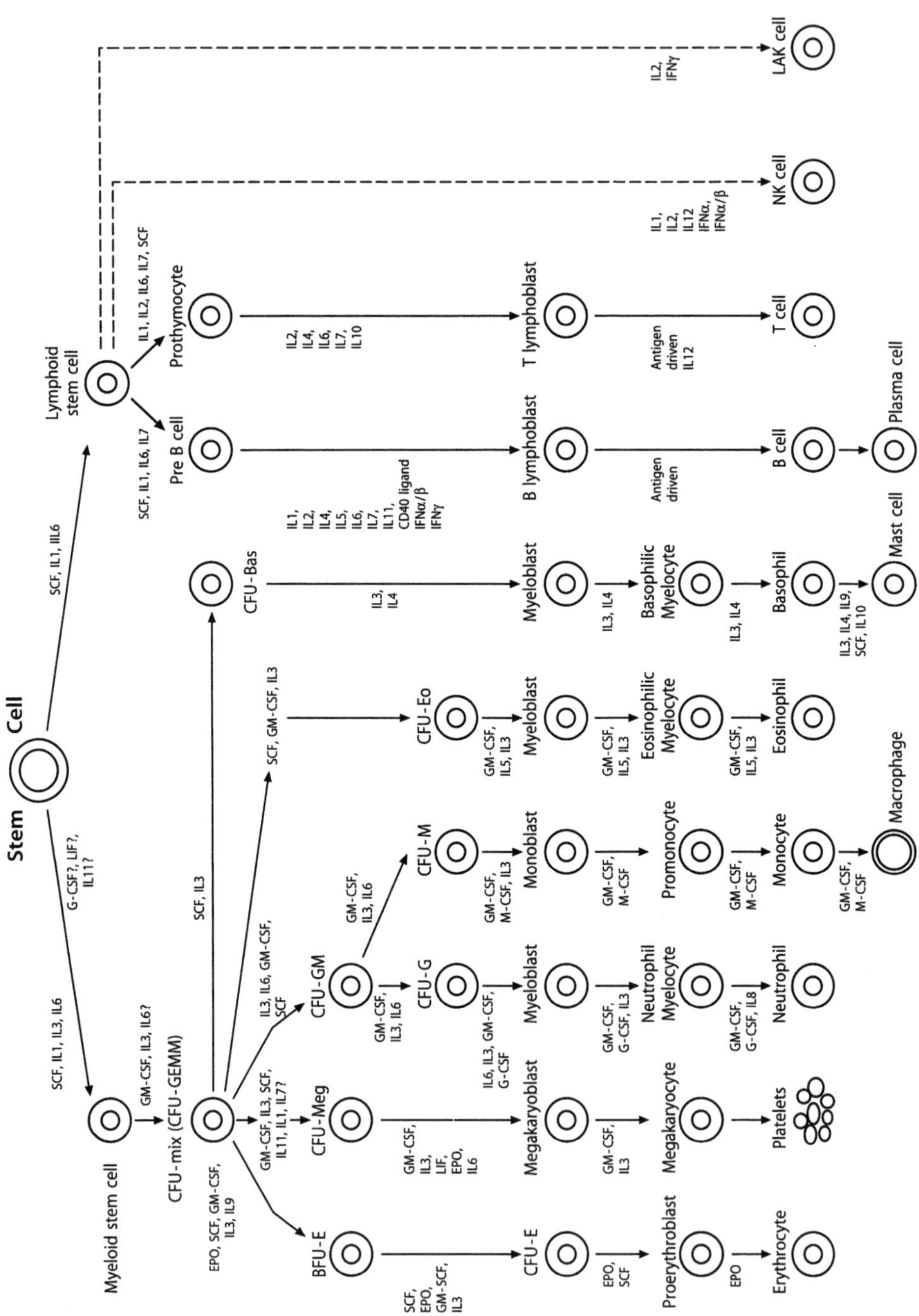

Fig. 2. The human haematopoietic system and its control by regulatory glycoproteins: stimulation of proliferation; differentiation and function are included. Adapted by permission from K. Atkinson, editor of Clinical bone-marrow transplantation: A Reference Textbook. Cambridge University Press. Cambridge 1994

Table 7. The haematopoietic growth factors

Function	Factors
Colony-stimulating factors	G-CSF GM-CSF M-CSF IL-3 IL-5 Erythropoietin Thrombopoietin
Stem-cell factors	Stem-cell factors Flt 3 ligand
Synergistic factors	IL-1 IL-6 IL-7 IL-9 IL-10 IL-11 IL-12 LIF
Inhibitors/bi-directional regulators	TNFα TGFβ MIP1α IFNδ

CSF, colony-stimulating factor; G, granulocyte; GM, granulocyte-macrophage; M, macrophage; IL, interleukin; TNF, tumour necrosis factor; TGF, transforming growth factor; MIP, macrophage inflammatory protein; IFN, interferon; LIF, leukaemia inhibitory factor.

3.5 Haematopoietic Recovery Post-transplant

Leucocytes usually begin to reappear in the blood during the second or third weeks post-transplant. The rate of reappearance of leucocytes, neutrophils, lymphocytes, monocytes, platelets and reticulocytes is shown in Fig. 3. Bone-marrow cellularity is normally decreased for the first 2–3 months post-transplant and may show two lineages at day 14 post-transplant, although all three cell lineages are usually represented in marrow aspirate samples by day 21 post-transplant.

3.6 Immunological Recovery Post-transplant

3.6.1 T Lymphocyte Number and Function

T-cell repopulation is abnormal in recipients early after transplantation and in long-term patients with chronic GVHD, and is manifested by low levels of $CD4^+$ cells and normal or high levels of $CD8^+$ cells, producing a reversal of the normal

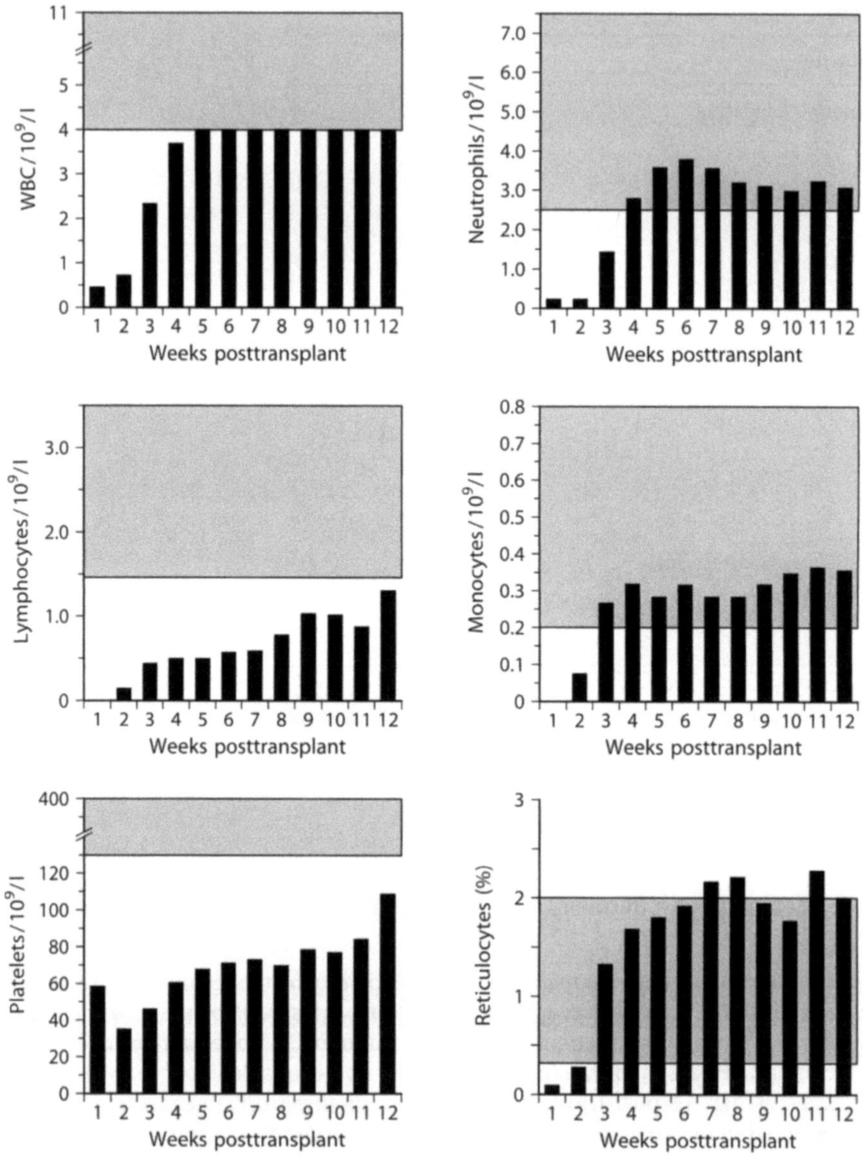

Fig. 3. Reconstitution of leucocytes, neutrophils, lymphocytes, monocytes, platelets and reticulocytes after human leucocyte antigen (HLA)-identical sibling marrow transplantation. The data are derived from 79 recipients of HLA-identical sibling marrow transplants for haematological malignancy, who were treated pre-transplant with cyclophosphamide 120 mg/kg and fractionated total body irradiation 12–15 Gy and were immune suppressed with cyclosporin. They received non-T cell-depleted bone marrow. Reproduced with permission, from Atkinson K. (1990) Bone Marrow Transplantation, 5: 209–226

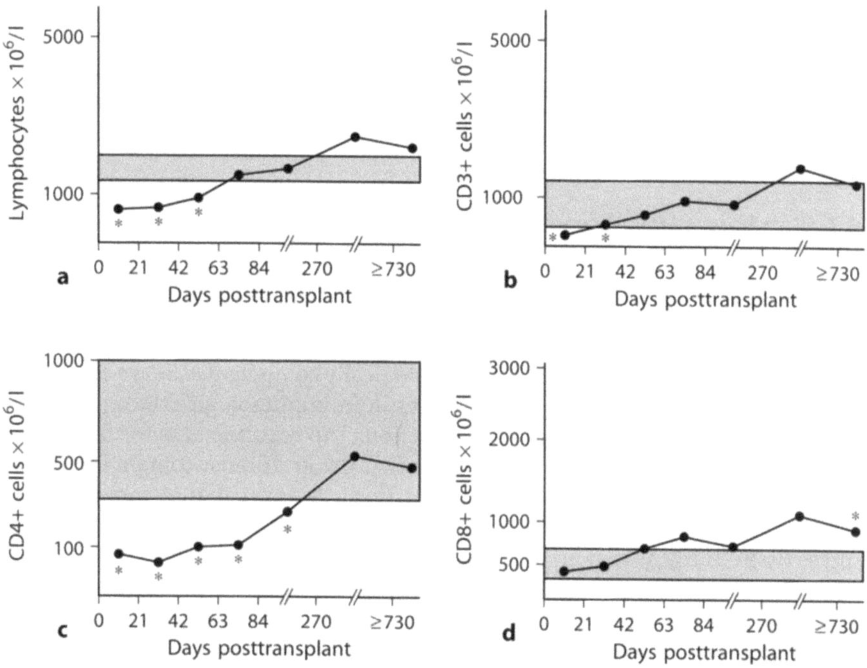

Fig. 4. Reconstitution of lymphocytes, total T cells (CD3+), CD4+ T cells and CD8+ cells. The data are derived from 63 recipients of non-T cell-depleted HLA-identical sibling marrow. Immune suppression post-transplant was with cyclosporin ($n = 53$) or methotrexate ($n = 10$). Values shown are mean values. *Asterisk* (*) represents values significantly different from normal. Reproduced, with permission from Atkinson K. (1990). Bone Marrow Transplantation 5: 209–226

CD4:CD8 ratio (Fig. 4). In contrast, large granular lymphocytes with natural killer activity constitute a major proportion of lymphoid cells repopulating the peripheral blood early after transplant.

Cellular immunity measured in vivo by skin testing, with delayed-type hypersensitivity-recall antigens, such as Candida, Mumps and Tricophytin, is impaired for as long as 2 years after allogeneic transplantation. Monocyte numbers in the blood return to normal rapidly post-transplant.

3.6.2 B Lymphocyte Number and Function

Prior to 3 months post-allografting, cell populations of immature B cells expressing CD19 and CD5 can be found in the blood. These immature B cells are not capable of producing quantitatively or qualitatively normal immunoglobulin (Ig) subclasses. By 3 months post-transplant, B lymphocytes have repopulated the peripheral blood in allogeneic, syngeneic and autologous marrow recipients.

3.6.3 Serum Immunoglobulin (Ig) Levels

Within the first 3 months after transplantation, serum Ig levels are low. By 1 year, serum IgG and IgM have reached normal levels, but IgA levels remain low until approximately 2 years post-transplant.

3.6.4 Graft-Versus-Host Disease (GVHD) and Graft–Host Tolerance

GVHD is a major complication of allogeneic bone-marrow or blood stem-cell transplantation. The transfer of tissues between normal individuals usually results in the recognition and destruction of the foreign tissue in a host-versus-graft reaction (graft rejection). However, if immunologically competent cells are contained in the transplanted graft, the transfer can result in immunological recognition in the other direction, a graft-versus-host reaction. The requirements for the development of GVHD include the requirements that the graft must contain immunologically competent cells, that the recipient must be incapable of mounting an effective response to destroy the transplanted cells, and that the recipient must express tissue antigens that are not present in the transplant donor.

In allogeneic bone-marrow or blood transplantation, the first requirement for a GVHD reaction is the presence of mature T cells in the marrow innoculum. The second requirement is the expression of recipient tissue antigens not present in the donor. In transplants between HLA-matched individual-donor recipient pairs, these antigens are, by definition, minor [non-MHC (non-major histocompatability complex)] histocompatibility antigens.

Without prophylactic immune suppression, allogeneic transplant recipient transplantation is complicated by GVHD. Acute GVHD can occur within days or as late as 2–3 months after transplant. Chronic GVHD is defined as that occurring 100 days or later post-transplant.

The immunopathophysiology of GVHD is currently understood as a process with two consecutive phases. Recipient tissues first activate donor T lymphocytes (afferent arm). These activated T cells then secrete cytokines, which recruit additional cells, induce the expression of histocompatibility antigens and focus the attack of donor effector cells on recipient targets (efferent arm) [8].

The most commonly utilized immune-suppressive regimen for recipients of T-replete (as opposed to T depleted) marrow grafts is the combination of cyclosporin and methotrexate. This normally needs to be given from immediately pre-transplant until between 6 months and 12 months post-transplant. In those patients who develop no or only mild chronic GVHD, cyclosporin immune suppression can be tapered and stopped at this time, since graft–host tolerance would have developed in such an individual.

There are three ways in which tolerance may be achieved: clonal deletion, clonal anergy, and active suppression. Evidence for all three mechanisms have been demonstrated in clinical and experimental bone-marrow transplantation.

4 Practical Considerations for Clinical Transplantation

4.1 Commonly Used Age Limits for Transplantation

Age limits for transplantation are often taken as 65 years for autologous blood stem-cell/marrow transplantation, 55 years for HLA-identical sibling transplantation and 50 years for unrelated donor transplantation.

Table 8. General pre-transplant work-up

Full blood count
Urea, electrolytes, creatinine, calcium, phosphate, uric acid, blood sugar Coagulation tests (PT/APTT)
Serology for CMV, HSV, VZV, HIV 1 and 2, HBV, HCV, EBV (VCA, EBNA, EBV, IgM), toxoplasma, HTLV 1 and 2 as appropriate
ABO/Rh typing
HLA typing/mixed lymphocyte typing (as appropriate)/lymphocyte crossmatch
Creatinine clearance rate (24-h urine collection or isotope scan)
CXR, ECG, left ventricular ejection fraction measurement
Microbiology screening
Throat swab, nose swab, Hickman catheter exit site swab (to detect organisms such as MRSA, MRSE, pseudonomas)
Cerebro-spinal fluid examination
In ALL, aggressive NHL or lymphoid transformation of CML
Of doubtful value in AML
Bone-marrow examination (morphology, cytogenetics, MRD determination and immune phenotyping)
Review of original pathology
Determination of chemosensitivity of underlying disease (as appropriate)
Autologous stem-cell cryopreservation (as appropriate)
Platelet support strategy (as appropriate)
Dental check
Social work department assessment
Sperm banking/embryo storage/IVF counseling
Remove any foreign bodies (e.g., biliary stents) if possible
Family conference and consent form signing
Start prophylactic medications (e.g., cotrimoxazole)

PT, prothrombin time; APTT, activated partial thromboplastin time; HSV, herpes simplex virus; VZU, varicella-zoster virus; HIV, human immune deficiency virus; HBV, hepatitis B virus; HCV, hepatitis C virus; EBV, Epstein-Barr virus; VCA, viral capsid antigen; BNA, Epstein-Barr nuclear antigen; HTLV, human T lymphotropic virus; CXR, chest x-ray; ECG, electrocardiogram; MRSA, multi-resistant *Staphylococcus aureus*; MRSE, multi-resistant *Staphylococcus epidermidis*; MRD, minimal residual disease; IVF, in vitro fertilization.

4.2 General Pre-transplant Work-Up

The general pre-transplant work-up schedule is conducted in the framework of a protocol as shown in Table 8.

4.3 Commonly Used Conditioning Regimens

Commonly used conditioning regimens for both autologous and allogeneic transplantation are shown in Table 9. The side effects of chemotherapy-based conditioning regimens include nausea, vomiting, diarrhoea, oropharyngeal mucositis, lethargy, alopecia, pancytopenia, haemorrhagic cystitis, hepatic veno-occlusive disease (VOD), interstitial pneumonitis, and secondary malignancy. Side effects of total body irradiation, in addition to the above, include acute parotitis, skin erythema, cataracts and (in children) growth retardation.

The routine tests often performed during the inpatient course of bone-marrow transplantation treatment are shown in Table 10, and a summary of the advantages and disadvantages of blood stem-cell transplantation when compared with marrow transplantation are shown in Table 11.

4.4 Methods of Mobilizing Peripheral Blood Stem Cells

In normal donors, current practice is to utilize a HGF alone, normally G-CSF at a dose of 10 μg/kg/day. Blood stem cells are usually predictably mobilized at day 5 of G-CSF administration and sufficient quantities (equal to or greater than 2×10^6 $CD34^+$ cells/kg) can often be collected in 1–2 daily leukaphereses.

Table 9. Commonly used pre-transplant conditioning regimens

	Regimen	Total dose administered
Autologous transplantation		
Lymphoma	CBV	Cyclophosphamide 4.8–7.2 gm/m^2 Carmustine 300–600 mg/m^2 Etoposide 750–2400 mg/m^2
	BEAC	Carmustine 300 mg/m^2 Etoposide 600–800 mg/m^2 Cytarabine 800 mg/m^2 Cyclophosphamide 140 mg/kg or 6 gm/m^2
	BEAM	Carmustine 300 mg/m^2 Etoposide 400–800 mg/m^2 Cytarabine 800–1600 mg/m^2 Melphalan 140 mg/m^2
	CY-TBI	Cyclophosphamide 120–200 mg/kg Total body irradiation 800–1320 cGy

Table 9 (continued)

	Regimen	Total dose administered
Autologous transplantation		
	VP16, CY, TBI	Etoposide 60 mg/kg or 750 mg/m^2 Cyclophosphamide 100 - 120 mg/kg Total body irradiation 1200 - 1375 cGy
Leukaemia	BUS-CY$_2$ Regimen	Busulphan 4 mg/kg/day for 4 days Total dose administered Cyclophosphamide 60 mg/kg/day for 2 days
	BUS-MEL	Busulphan 4 mg/kg/day for 4 day Melphalan 140 mg/m^2 once Cyclophosphamide 60 mg/kg/day for 2 days
	CY-TBI	Cyclophosphamide 60 mg/kg/day for 2 days Total body irradiation 800 - 1200 cGY
Myeloma	Melphalan	Melphalan 100 mg/kg/m^2/day for 2 days
	Melphalan/TBI	Melphalan 140 mg/m^2/day once Total body irradiation 850 cGy
Breast Cancer	CTC (STAMP V)	Cyclophosphamide 1.5 g/m^2 IV daily for 4 days Thiotepa 125 mg/m^2 IV daily for 4 days Carboplatin 200 mg/m^2 IV daily for 4 days
Allogeneic transplantation		
HLA-identical sibling transplant Haematological malignancy	Bus-Cy$_2$	Busulphan 4 mg/kg/day × 3.5 - 4 days[a] Cyclophosphamide 60 mg/kg/day × 2 days
	CY-TBI[b]	Cyclophosphamide 60 mg/kg/day × 2 days Total body irradiation 10 - 13 Gy: one common protocol utilizes TBI 2 Gy twice daily for 3 days or 2 Gy once daily for 6 days
	Regimen	Total dose administered twice daily for 6 days
Severe aplastic anaemia	CY-ATG	Cyclophosphamide 50 mg/kg/day (4 doses) alternating with anti- thymocyte globulin (Upjohn) 15 mg/kg/day (3 doses)

[a] Busulphan 14 mg/kg total dose often used for myeloma; 16 mg/kg total dose usually used for other haematological malignancies.

[b] A CY-TBI regimen may be used in preference to a chemotherapy only regimen in patients with CNS involvement, prior exposure to family member blood products, and CLL/NHL.

Table 10. Routine tests on bone-marrow transplant recipients (while an in-patient)

Test	Frequency
Full blood count, urea, sodium, potassium, creatinine, glucose	Daily
Liver function tests, calcium, magnesium, phosphate	Monday, Wednesday, Friday
PT, APTT	Tuesday and Friday
Group and hold	Tuesday and Friday
Antibiotic levels	Monday, Wednesday, Friday (as appropriate)
Nose, throat, catheter exit site swabs, MSU	Weekly
Patients on TPN – calcium, magnesium phosphate	Daily
Chest X-ray	Weekly

MSU, midstream specimen of urine; TPN, total parenteral nutrition.

Table 11. Advantages and disadvantages of blood stem-cell transplantation compared with marrow transplantation

Advantages	Disadvantages
To recipient	
Faster neutrophil recovery	More GVHD? (allo transplants)
Faster platelet recovery	
Faster immunological recovery	
Less IV antibiotics	
Less fever = 38.5 °C	
Shorter hospitalization	
Lower cost	
More GVL? (allo transplants)	
To donor	
No general anaesthetic	Venepuncture or central IV catheter for leukapheresis
No marrow harvest	Side effects from G-CSF
No hospitalization	

The same strategy can be utilized for harvesting autologous peripheral blood stem cells; however, such stem cells are less readily harvested in patients who have had six or more courses of prior conventional-dose cytotoxic chemotherapy, with or without local radiotherapy. For this reason, a chemotherapeutic agent, often cyclophosphamide or a disease-specific chemotherapy regimen, is utilized in conjunction with G-CSF.

The best known of such regimens includes cyclophosphamide 1.5–4 g/m^2 on day 1, followed by G-CSF 10 mg/kg per day on days 2–12, with leukapheresis being performed usually on days 10–12, starting when the white count reaches 1.0×10^9/l on recovery after the cyclophosphamide-induced nadir.

The yield of CD34$^+$ progenitor/stem cells in the leukapheresis harvest can be predicted by monitoring the peripheral blood CD34$^+$ cell count. It is not usually

Table 12. Disease-free survival after autologous bone-marrow or blood stem-cell transplantation

Disease	Status at time of transplant	Approximate 5-year disease-free survival (%)
AML	CR1	50
AML	≥ CR2	20–40
AML	Not in remission	10–20
ALL	CR1	30–50
ALL	≥ CR2	10–30
NHL	Low grade CR1	60–80
NHL	Low grade CR2/first relapse	40–60
NHL	Intermediate/high grade CR1	60–80
NHL	Intermediate graft first relapse	30–40
NHL	Intermediate grade chemosensitive relapse	30–40
HD	First relapse	60
Myeloma		30
Breast cancer	Stage II	70
Breast cancer	Stave III	60
Breast cancer	Stage IV	10–40
Non-seminomatous germ cell tumour of testis	Relapsed or chemorefractory	30–40

CR, complete remission.

worth leukapheresing a patient if the blood $CD34^+$ cell number is less than 10/µl (0.01×10^9/l). A good yield in the apheresis harvest is obtained if the $CD34^+$ cell count is at least 50/µl.

5 Disease-Free Survival After Autologous Transplantation

Outcome depends markedly on the disease being treated and the stage of the disease when the transplant is performed. Generally, transplantation earlier in the course of the disease results in superior results than that performed. Additionally, if patients are resistant to conventional-dose chemotherapy, they will in general respond poorly to high-dose chemotherapy.

Approximate 5-year survival figures are shown in Table 12 for the diseases for which autologous transplantation (marrow or blood stem cell) is commonly performed.

6 Results of HLA-Identical Sibling Transplantation

Similar disease-free survival data are shown for HLA-identical sibling transplantation in Table 13.

Table 13. Disease-free survival after HLA-identical sibling bone-marrow transplantation

Disease	Status at time of transplant	Approximate 5-year disease-free survival (%)
AML	CR1	60
AML =	≥ CR2	20–40
AML	Not in remission	20
ALL	CR1	50
ALL	≥ CR2	20–40
ALL	Not in remission	10–20
CML	CP1	60–80
CML	Acceleration	20–40
CML	BT	10–20
CLL		40–50
MDS	RA/RAS	50
MDS	RAEB	30
MDS	RAEBt	30
MDS	AML	20
Myeloma		30–40
SAA		80–90
Thalassaemia major		65–95
Sickle cell disease		70–80

CP, chronic phase; BT, blast transformation; RA, refractory anaemia; RAS, refractory anaemia with ring sideroblasts; RAEB, refractory anaemia with excess blasts; RAEBt, refractory anaemia with excess blasts in transformation.

7 Clinical Problems in Bone-Marrow Transplantation

A list of transplant-related complications is given in Table 14. Two of the more common complications are dealt with in Sect. 7.1 and Sect. 7.2. For others, further reading is recommended at the end of this review.

7.1 Febrile Neutropenia

Febrile neutropenia is a common complication during the first several weeks post-transplant, before neutrophil recovery has occurred. Neutropenia is defined as an absolute neutrophil count of less than 1.0×10^9/l, and significant fever is usually taken as being equal to or greater than 38.5 °C. All such fevers should be assumed due to infection until proven otherwise. When they occur, cultures should be taken of blood, urine, throat, intravenous-catheter exit site and any other potential sites of sepsis, and broad-spectrum, intravenous anti-bacterial antibiotics should be commenced immediately. Subsequent management depends on the response to the initial antibiotic treatment and is outlined in Table 15.

For persistent fever or fever occurring in the presence of an adequate neutrophil count, the possible causes include: cryptic bacterial infection (for example sinusitis), central venous-catheter infection, fungaemia, drug fever (including antibiotic fever), and viraemia [especially due to cytomegalovirus (CMV)]

Table 14. Transplant-related complications

Failure of sustained engraftment
Febrile neutropenia
Fungal infections
Viral infections
Acute graft-versus-host disease
Chronic graft-versus-host disease
Interstitial pneumonitis
Hepatic VOD
Haemorrhagic cystitis
Cyclosporin-related hypertension
Retarded growth and development
New malignancy after BMT

BMT, bone-marrow transplant; VOD, veno-occlusive disease.

Table 15. Management of fever of unknown origin in neutropenic patients after marrow transplantation

Time from development of fever (h)	Strategy
0	1. Rapid assessment (physical exam) and culture (blood, urine, throat, central venous catheter exit site). Chest X-ray
0	2. Start IV broad-spectrum antibacterial antibiotics[a]
48	3. Add IV vancomycin
72	4. Add antifungal antibiotic
96	5. Remove indwelling catheter
	6. Additional measures if severe infection. i) G-CSF or GM-CSF ii) Renal dose dopamine/packed cell transfusion (if blood pressure compromised)

[a] Follow order of treatment if no response.

7.2 Viral Infections

The most common viral infections are those caused by DNA herpes viruses, particularly herpes simplex virus (HSV), CMV and varicella-zoster virus (VZV). They occur considerably more frequently after allogeneic than autologous transplantation, and the time frame for their occurrence post-transplant differs and is shown in Fig. 5. Diagnostic methods for each of them are described in Table 16.

It is common practice to administer routine prophylaxis with acyclovir for HSV and with ganciclovir for CMV. Such an approach has markedly diminished the morbidity from oropharyngeal mucositis due to HSV and the morbidity and mortality from interstitial pneumonitis due to CMV.

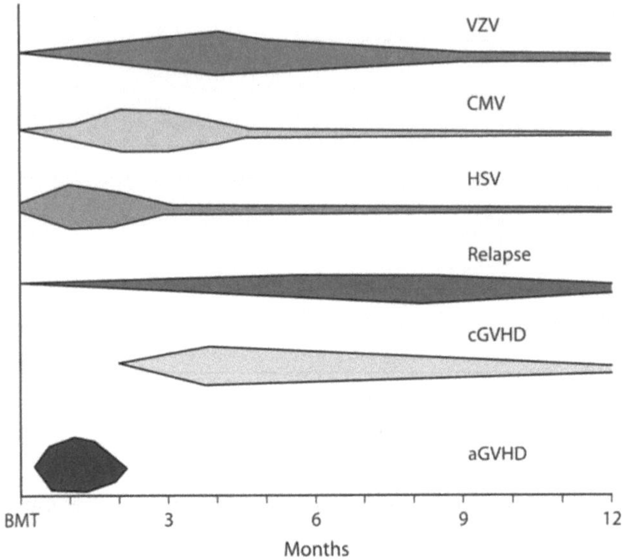

Fig. 5. Timeframe of common DNA herpes virus infections after bone-marrow transplanation (BMT). Adapted with permission from Clinical Bone Marrow Transplantation: A Reference Textbook. Ed K. Atkinson. Cambridge University Press. Cambridge 1994

Table 16. Diagnosis of DNA herpes virus infection

Virus	Diagnostic test
CMV	CMV culture (including use of shell vial technique)[a] Histology of tissue biopsy[b] Monoclonal antibodies (linked to enzymes) against CMV antigens (tissue sections)[b] In situ hybridization using radiolabeled DNA probe on tissue sections Serology Restriction enzyme analysis of viral DNA (RFLP) Polymerase chain reaction amplication of viral DNA
HSV	HSV culture, histology of tissue biopsy[b], serology Monoclonal antibodies against HSV antigens (vesicle[c]/ulcer smear[c], or tissue sections)[b]
VZV	VZV culture, histology of tissue biopsy[b], serology Monoclonal antibodies against VZV antigens (vesicle[c]/ulcer smear[c], or tissue sections)[b]

[a] 24 h.
[b] 6–48 h.
[c] 3 h.
DNA, deoxyribose nucleic acid; RFLP, restriction fragment length polymorphism.

8 Acute GVHD

Risk factors for the development of acute GVHD include histoincompatibility, allosensitization of the donor by prior pregnancy or prior blood transfusion, increasing donor age, increasing patient age, gender mismatch between donor and recipient (especially female donors for male recipients), omission of GVHD prophylaxis, infusion of viable donor leucocytes, and insufficient intensity of the conditioning regimen. The most common target organs of acute GVHD are the skin, liver and gut but the immune system itself is functionally impaired by GVHD. A clinical staging system for the severity of skin, liver and gut involvement has been established, with the overall performance status contributing to the overall grade (I–IV, mild through severe) of acute GVHD, and a histopathological grading system has also been developed for these three organs and (see Table 17).

Treatment of acute GVHD is usually with prednisone 2 μg/kg per day initially, with continuation of cyclosporin prophylaxis. Higher doses of prednisone (up to 1 g intravenously, daily for 3 days) may be used for severe acute GVHD. A third agent is anti-thymocyte globulin, which should usually be used after or during corticosteroid withdrawal to minimize infectious toxicity.

While patients with grade I–II GVHD usually recover, those with grade IV acute GVHD usually do not. Conventional immune prophylaxis with cyclosporin/methotrexate usually prevents severe acute GVHD after HLA-identical sibling transplantation, but not always after HLA-identical unrelated donor transplantation.

Table 17. Histopathological grading system (adapted from [9])

Grade	Description
Acute cutaneous GVHD	
Grade I	Vacuolar degeneration of basal epidermal cells
Grade II	Grade I plus scattered necrosis of epidermal cells with "eosinophilic" (colloid) bodies and spongiosis
Grade III	Grade II plus focal dermo-epidermal separation and cleft formation
Grade IV	Extensive desquamation
Acute gastrointestinal GVHD	
Grade I	Single-cell necrosis of crypt-gland epithelium with focal dilatation of glands
Grade II	Grade I plus focal or diffuse loss of glands/crypts
Grade III	Grade II plus focal mucosal ulceration
Grade IV	Diffuse ulceration
Acute hepatic GVHD	
Grade I	Less than 25% of small bile ducts showing degenerate epithelium ± necrosis
Grade II	25%–49% of small bile ducts involved
Grade III	50%–74% of small bile ducts involved
Grade IV	Greater than or equal to 75% of small bile ducts involved

9 Chronic GVHD

Chronic GVHD is defined as that occurring 100 days or later after the transplant. It may develop de novo, without any preceding acute GVHD, progressively where acute GVHD merges into chronic GVHD, or sequentially where there is a GVHD-free interval between acute and chronic GVHD. It may either be limited in extent the skin or liver, or both, are involved or extensive if any other organ involvement occurs. The incidence after HLA-identical sibling bone-marrow transplantation is approximately 50% and usually has resolved by 4–5 years post-transplant, although occasional cases can last for 12 years and longer. The peak incidence is between 1 year and 2 years post-transplant. The most common organs involved are skin, mouth, liver, eyes, small intestine and lung, and risk factors include prior acute GVHD, increasing patient age, increasing recipient age, HLA-partially-identical family-member transplants (as opposed to HLA-identical family-member transplants), HLA-identical or partially identical unrelated donor transplants, infusion of viable donor leucocytes (buffy-coat infusions) and second marrow transplants.

Chronic GVHD leads to severe humoral and cellular immune deficiency which, in turn, leads to a markedly increased risk of late infections. The most common infections are bacterial, usually caused by a gram-positive cocci and Haemophilus influenzae, but fungal and viral infections also are not uncommon.

Clinical manifestations differ from those of acute GVHD, particularly with regard to oral and ocular involvement, where a sicca syndrome can cause dry eyes and dry mouth. Severe chronic skin changes can occur with dyspigmentation, sclerodermatous subcutaneous fibrosis with development of hide-bound and sometimes ulcerated skin, nail dystrophy and alopecia. The siccus syndrome can also involve the vagina and an obliterative bronchiolitis syndrome occurs, as does thrombocytopenia. The latter is a bad prognostic indicator.

The hallmark of treatment is combined predisone and cyclosporin immune suppression, which is often required for up to 9 months. Thalidomide has some utility, as does psoralen and ultraviolet A (PUVA phototherapy) for skin or oral involvement.

10 Interstitial Pneumonitis

Interstitial pneumonitis (IP) used to occur in approximately 20% of recipients of allogeneic transplants prior to the introduction of prophylactic ganciclovir. This, however, has basically eradicated early cytomegalovirus interstitial pneumonitis, and the most common cause of interstitial pneumonitis post-transplant now is "idiopathic" IP, thought most commonly to be due to the pre-transplant chemotherapy or chemoradiotherapy regimen. Pneumocystis carinii pneumonitis is also now rare, with the widespread use of prophylactic cotrimoxazale both pre- and post-transplant. Rare miscellaneous causes of IP include fungal infections, Chlamydia trachomatis, herpes virus type-6 infection, adenovirus infection, and infection with legionella, mycoplasama, respiratory syncytial virus, influenza virus, parainfluenza virus and toxoplasma.

Table 18. Protocol for processing open lung/transbronchial biopsy specimens for diagnosis of interstitial pneumonitis

Histopathology	Microbiology
1. Haematoxylin and eosin stain	1. Gram stain (impression slide of a cut edge)
2. CMV early antigen immunoperoxidase stain	2. ZN Stain
3. Methenamine silver stain for pneumocystis	3. Legionella direct fluorescence stain
4. PAS stain for fungi	4. Pneumocystis indirect fluorescence stain (make impression slide and send one to histopathology laboratory fixed in formalin)
	5. Bacterial culture including aerobes, anaerobes, AFB, legionella, nocardia
	6. Fungal culture
	7. Viral culture (use shell vial technique) and chlamydial culture. (After normal working hours this sample may be put into viral culture medium in a Nunc tube (freezing-resistant) and placed in liquid nitrogen container)

Divide sample/s aseptically. PAS, periodic acid-Schiff; ZN, Ziehl-Nielsen; AFB, ACID-fast bacillus.

The classical clinical syndrome of interstitial pneumonitis is one of fever, dyspnoea and tachypnea, with very little in the way of audible abnormal chest signs and a diffuse ground-glass appearance not restricted to anatomical borders on chest X-ray. This contrasts with the air-space consolidation restricted to anatomical borders in lobar pneumonia or the bilateral basal patchy consolidation in bronchopneumonia.

Diagnosis usually requires a tissue biopsy either by the transbronchial route or by open-lung biopsy. A protocol for processing such lung-tissue samples is shown in Table 18.

11 Hepatic Veno-occlusive Disease

Hepatic veno-occlusive disease (VOD) usually occurs in the first 3–6 weeks post-transplant and is caused by damage to the hepatic venules by the high-dose pre-transplant chemotherapy or chemoradiation conditioning regimen. The diagnostic triad of jaundice, ascites and weight gain suggests the diagnosis. Right-sided upper abdominal pain (due to stretching of the hepatic capsule), pleural effusions, oliguria and sudden marked thrombocytopenia also occur. The differential diagnosis includes hepatic-vein obstruction, pericardial effusion, pancreatic disease, peritonitis and occult cirrhosis with hypoalbuminaemia.

Liver function abnormalities are similar to those seen in acute hepatic GVHD with conjugated hyperbilirubinaemia, a marked (often sixfold or greater) elevation of serum alkaline phosphatase and a more modest elevation of hepatic transaminases. Risk factors include pre-transplant conditioning with busulphan and cyclophosphamide, and prior liver disease. It is thought that low-dose intravenous heparin, started pre-transplant at 100 units/kg for 24 h, may decrease the incidence. The disease varies in severity but severe hepatic VOD is usually fatal.

Management of VOD is essentially supportive with maintenance of the intravascular volume and renal perfusion as the primary aim. Thrombolytic therapy with recombinant human tissue plasminogen activator 5–10 mg intravenously, daily for up to 6 days, has been utilized, but has yet to be shown to be definitively beneficial.

12 Role of the Pathologist in Bone-Marrow Transplantation

12.1 The Haematology and Blood Bank Laboratory

The provision of daily or twice-daily, full blood counts during the in-patient transplant procedure, as well as twice-weekly coagulation profiles, is an integral component of the monitoring process during marrow transplantation. Likewise, the provision of blood products, primarily packed cells and platelets during the period of haematopoietic reconstitution is a fundamental requirement.

While major or minor blood type ABO incompatibility between donor and recipient is not a contraindication to transplantation, both need to be carefully managed. The prophylaxis of alloimmune haemolysis due to major ABO incompatibility includes removal of the red cells from the marrow innoculum, for example, by automated cell separation. Additionally, the titer of isoagglutinin in recipient plasma must be reduced, for example, by plasma exchange.

Another responsibility of the laboratory haematologist in the bone-marrow-transplant program is the cryopreservation and functional assessment of harvested autologous bone marrow or blood stem cells. Preparation of these for cryopreservation includes the need for volume reduction, addition of cryoprotectant, controlled rate freezing and storage in the liquid phase (–196°C) or in the vapor phase (–156°C) of liquid nitrogen. Assessment of the frozen product includes analysis of the $CD34^+$ cell content by flow cytometry and CFU-GM content by agar culture.

12.2 Microbiology Laboratory

The microbiology laboratory is clearly a key service in support of a marrow-transplant/stem cell-transplant program, with responsibility for identification of the bacterial, fungal and viral pathogens, during the period of impaired immunity experienced by transplant recipients during the first several weeks and months post-transplant. One of the most important functions is the processing of lung tissue in patients with interstitial pneumonitis in an attempt to determine the cause (Table 18).

12.3 Histopathology Laboratory

A third key laboratory support to a marrow-transplant program is the histopathology laboratory, most commonly for the diagnosis of GVHD on tissue sections, but also for the diagnosis of opportunistic infections, particularly in the lung. The histopathological grading system for skin, liver and gut involved by acute GVHD is described in Table 17, and the reader is referred to more extensive texts at the end of this review (Sect. 13.1) on the appearance of tissues involved by acute or chronic GVHD.

12.4 Cytogenetics Laboratory

Since the objective of marrow or blood stem-cell transplantation is cure, the detection of minimal residual disease post-transplant is assuming increasing importance. One way of detecting this in transplantation for malignant disorders is to search for the presence of a cytogenetic abnormality present pre-transplant. Examples of some numerical and structural cytogenetic abnormalities commonly associated with haematological malignancies is given in Table 19. Some of these

Table 19. Some numerical and structural cytogenetic abnormalities commonly associated with haematological malignancies

Disease	Subclassification	Reported abnormalities
ANL	M1 (myeloblastic without differentiation)	+8t(9;22) (q34;q11)
	M2 (myeloblastic with differentiation)	t(8;21) (q22;q22) often associated with loss of a sex chromosome
	M3 (promyelocytic)	t(15;17) (q22;q12)
	M4 (myelomonocytic)	+4 +8
	M4Eo (myelomonocytic with eosinophilia in bone marrow)	inv(16) (p13;q22), t(16;16)(p13;q22), del(16) (q22)
	M5 (monocytic); more often in M5a (without differentiation)	t(9;11) (p22;q23), 11q23 abnormalities, +8
	M6	Complex rearrangements, del(5q), +8
	M7 (megakaryoblastic)	Complex rearrangements involving –5 or del(5q), –7 or del (7q)+8+9t(1;22)+21
ALL	pre-B lineage	t(4;11) (q21;q23)t(1;19) (q23;p13)t(9;22) (q34;q11)

Table 19 (continued)

Disease	Subclassification	Reported abnormalities
ALL	B lineage	>50 chromosomes/cell most commonly +X, +4, +6, +10, +14, +17, +18, +20, +21
	L3 (Burkitt's) B lineage	t(8;14) (q24;q32)t(2;8) (p12;q24)t(8;22) (q24;q11)
	B and T lineage	del(6q)del(11) (q23)
	T lineage	inv(14) (q11;q32)t(8;14) (q24;q11)t(10;14) (q24;q11)t(11;14) (p13;q11)7q32 → 7q36 rearrangements
CML	Chronic phase	t(9;22) (q34;q11)
	Accelerated or blastic phase	Abnormalities additional to t(9;22) include +8, i(17q), +19, +der(22)t(9;22)
CLL	B cell	+1214q32 abnormalities e.g. t(11;14) (q13;q32) t(14;18) (q32;q11) t(2;14) (p13;q32) del(6q) del(11q) del(12p) 13q abnormalities
	T cell	inv or t(14) (q11)t(11;14) (p13;q11)
Myelo-dysplasia	RA	del(5q)+8
	RAS	del(5q)+8del (20q)
	RAEBt	del(5q)–7 or del(7q)+812p abnormalitiesi(17q)
Myelo-proliferative disorder	Polycythaemia rubra vera	del(20) (q11)
Non-Hodgkin's	Follicular	t(14;18) (q32;q21)
Lymphoma	Diffuse (non-Burkitt's)Burkitt's	t(11;14) (q13;q32) t(8;14) (q24;q32) t(8;22) (q24;q11) t(2;8) (p12;q24)
	Additional abnormalities not specific for lymphoma type	del(6) (q21)i(6p)+12

can now be detected by molecular biological techniques and examples are shown in Table 20. As seen in Table 21, polymerase chain reaction (PCR) technology is 3–4 logs more sensitive than conventional cytogenetic technology. Other uses for cytogenetic or PCR evaluation of such abnormalities is the selection of patients at high risk of relapse after conventional therapy, as well as the assessment of the efficacy of marrow-purging protocols.

Table 20. Molecular markers amenable to detection by polymerase chain reaction

Cytogenetic abnormality	Molecular marker	Disease
None	CDRIII	B-lineage ALL, NHL
None	TCR	T-lineage ALL, NHL
t(14;18)	BCL2-IGH	85% FSCL, 25% DLCL
Deletion	SIL-TAL	25% T-lineage ALL
t(9;22)	BCR-ABL fusion	100% CML
t(1;19)	E2A-PBX1 fusion	25% pre-B ALL
t(15;17)	PML-RARα fusion	100% APML
t(6;9)	DEK-CAN fusion	AML-M2, M4
t(17;19)	E2A-HLF fusion	Rare ALL

Table 21. Assays for detection of minimal residual disease

Method	Marker	Sensitivity
Routine staining	Cellular morphology	10^{-1} to 10^{-2}
Cytogenetics	Chromosome morphology	10^{-1} to 10^{-2}
Fluorescence in situ hybridization	Chromosome structure	10^{-2}
Gene rearrangement	DNA configuration	10^{-2} to 10^{-3}
Flow cytometry	Antigen profile	10^{-3}
Clonogenic culture	In vitro growth	10^{-5}
Polymerase chain reaction	DNA/RNA structure	10^{-5}

12.5 Biochemistry Laboratory

Since marrow transplantation is basically the practice of internal medicine on a background of severe pancytopenia, it is not surprising that many serious medical problems occur in the early post-transplant period, including renal impairment, hepatic impairment, cardiac impairment and pulmonary impairment. Appropriate monitoring of the blood is essential in the daily management of each of these complications. The routine blood tests performed on marrow transplant in patients are shown in Table 10 and, clearly, additional investigations may be required according to the specific clinical problem present at the time.

12.6 Histocompatibility Typing Laboratory

HLA matching between donor and recipient is key to the success of allogeneic transplantation and this can only be supplied by a first-rate professional tissue-typing laboratory. An example of the inheritance of HLA types is shown in Fig. 6 and tissue-typing techniques currently used are detailed in Table 22.

				Mother		Father			
			HLA-A:	1	2	2	2		
			HLA-B:	8	44	39	27		
			HLA-DR:	3	2	7	1		
Spouse		Patient		Sib 1		Sib 2		Sib 3	
1	3	1	2	1	2	2	2	1	2
8	35	8	39	8	27	44	27	8	39
3	1	3	7	3	1	2	1	3	7
		Child 1		Child 2		Child 3			
		1	1	3	2	3	1		
		8	8	35	39	35	8		
		3	3	1	7	1	3		

Fig. 6. Inheritance of HLA-type. Adapted with permission from Clinical Bone Marrow Transplantation: A Reference Textbook. Ed K. Atkinson. Cambridge University Press. Cambridge 1994

Table 22. Currently used tissue-typing techniques

Technique	Comment
Serological Cellular	For routine HLA-A, -B, -DR typing
MLC) or MLR	Gives overview of HLA-D region compatibility
PLT	For HLA-DP typing
HTC	For HLA-Dw phenotyping: a panel of T cell defined, HLA-D region-associated antigens (HLA-Dw) are recognized. Primarily HLA-DR specific
CTLp	Appears to predict severity of acute GVHDafter T cell depleted unrelated donor transplantation
HTLp Biochemical	Currently being explored
IEF	Can distinguish variants of a given allele differing by a single amino acid
Molecular RFLP	Restriction enzymes digest genomic DNA into fragments of characteristic length, which are then separated by gel electrophoresis, blotted on to membranes and hybridized with radiolabelled cDNA probes, to reveal intron region polymorphisms for HLA-DR, -DQ and -DP genes. Most DR specificities can be distinguished by RFLP analysis, although some, especially of the DR4 family, cannot. Inferior to allele-specific oligonucleotide typing
SSO or ASO typing using PCR	DNA is amplified by PCR using oligonucleotide primers specific for nucleotide sequences within known hypervariable regions within the first domain of Class II alpha and beta chains. At least 17 alleles of DR4 have been identified by SSO-PCR typing. Currently it is the molecular method of choice

PCR, polymerase chain reaction; MLC, mixed lymphocyte culture; MLR, mixed lymphocyte reaction; PLT, primed lymphocyte test; HTC, homozygous typing cell testing; CTLp, cytotoxic T-cell precursor frequency analysis; HTLp, helper T-cell precursor frequency; IEF, isoelectric focusing; RFLP, restriction fragment length polymorphism; SSO, sequence-specific oligonucleotide; ASO, allele-specific oligonucleotide.

12.7 Immunology Laboratory

$CD34^+$ cell analysis is fundamental to assessment of the adequacy of the stem-cell dose harvested from autologous or allogeneic blood or marrow. Serum immunoglobulin and T and B cell and T cell subset analysis early post-transplant can be useful in evaluating the extent of the immune reconstitution. Such monitoring will prove even more important as, increasingly, techniques such as the administration of interleukin 2 are initiated to accelerate or augment immune recovery post-transplant.

13 New Developments

This field continues to be extremely active and a large number of new developments are currently underway. These are summarized in Table 23.

Clearly, the current era of dynamic evolution and expansion in this area of clinical medical science is increasing and major developments will occur in both cellular-component therapy and gene-transduction therapy over the next several years and well into the next century. The pathology laboratory will continue to play an integral and increasing role in this development.

Table 23. New developments

Development	Potential advantages
Use of purified stem cells	Autologous: minimization of tumour cell contamination Allogeneic: T-cell depletion
Ex vivo expansion of stem and progenitor cells	Bigger cell dose for transplant: minimization of neutropenia and thrombocytopenia.
New immunosuppressive agents, e.g., FK506, rapamycin, mycophenolate mofetil	Better control of GVHD
Adoptive cellular immunotherapy	Control of viral infections and viral-associated neoplasmas (e.g., EBV lymphoproliferative disease) Eradication of relapse of malignancy post-allograft
Transplantation for autoimmune disease	Possibility of cure for diseases such as rheumatoid arthritis, SLE, scleroderma, PAN
Gene transduction	A new form of therepy for:
TK (suicide) gene into T cells	Control of GVHD
Multi-drug resistance gene into HSC	Post-transplant chemotherapy without impaired blood counts
Rev M10 gene into HSC	Inactivation of HIV

13.1 Recommendations for Further Reading

1. Clinical bone-marrow transplantation (1994) In: K. Atkinson (ed) A reference textbook. Cambridge University Press, Cambridge
2. The BMT data book: a manual for bone-marrow and blood stem-cell transplantation (1997) K. Atkinson (ed) Cambridge University Press, Cambridge
3. Bone-marrow transplantation (1994) S. Forman, K. Blume, E.D. (eds) Thomas Blackwell Scientific Publications, Boston

Acknowledgements. I would like to thank Beth Lundberg for her skill and efficiency in preparing the manuscript. Tables 5, 15, 17, 18 and 21 were produced by permission from Kerry Atkinson, the editor of Clinical bone-marrow transplantation: a reference textbook (1994) Cambridge University Press. Tables 6, 7, 8, 10, 14, 19 and 20 were reproduced with permission from K. Atkinson, the editor of the BMT data book (1997) Cambridge University Press.

References

1. Thomas ED, Lochte HL, Lu WC, Ferrebee JW (1957) Intravenous infusion of bone marrow in patients receiving radiation and chemotherapy. New Engl J Med 25: 7491–7496
2. Mathé G, Amiel JL, Schwarzenberg L, et al. (1963) Hematopoietic chimera in man after allogeneic (homologous) bone marrow transplantation. BMJ 21: 633–635
3. Dausset J (1958) Iso-leuco-anticorps. Acta Haematologica 20: 156–166
4. Thomas ED, Buckner CD, Rudolph PH, et al. (1971) Allogeneic marrow grafting for hematological malignancy using HLA-matched donor-recipient sibling pairs. Blood 38: 267–287
5. Thomas ED, Buckner CD, Storb R, et al. (1972) Aplastic anemia treated by marrow transplantation. Lancet 1: 284–289
6. Thomas ED, Buckner CD, Banaji M, et al. (1977) One hundred patients with acute leukemia treated by chemotherapy, total body irradiation, and allogeneic marrow transplantation. Blood 49: 511–533
7. Thomas ED, Buckner CD, Clift RA, et al. (1979) Marrow transplantation for acute nonlymphoblastic leukemia in first remission. New Engl J Med 301: 597–599
8. Ferrara JL, Deeg HJ (1991) Graft versus-host-disease. New Engl J Med 324: 667–674
9. Lerner et al. (1974) Transplant Proc 6: 367–371

Pathology of Heart Transplant

J. P. MINDÁN and A. PANIZO

1 Introduction

Heart transplantation is a highly effective method for the treatment of patients with end-stage heart disease. Since Barnard performed the first clinical heart transplant in 1967, hundreds of centres around the world have been performing several thousands of heart transplants every year. With the success of this therapeutic technique, heart biopsies have had an important role, especially in the early diagnosis of most complications.

Right ventricular endomyocardial biopsy (EMB) has proven to be safe and appropriate for the management of heart allografts. The bioptome is introduced by trans-jugular vein to obtain three to six pieces of myocardium. EMB has a sampling error, but to date has prevailed as the most accurate method for diagnosing acute rejection [1], although some biopsies are non-informative as the material obtained is composed of fibrin, infiltrating fat or scar collagen. The complication rate of EMB is 1%–1.5% and includes ventricular extrasystoles or mild arrhythmias [2].

Current Topics in Pathology
Volume 92, C. L. Berry (Ed.)

Table 1. Most important morphological alterations of heart allografts

Hyperacute rejection
Early cardiac allograft failure (ECAF)
Preservation injury
Acute cellular rejection (ACR)
Previous biopsy sites
Quilty effect
Humoral rejection and acute vascular rejection
Infections
Neoplasias
Heart hypertrophy and fibrosis
Denervation
Calcification
Chronic rejection (CR)

This chapter emphasizes the role of EMB in monitoring heart transplants and describes the most important lesions diagnosed by the biopsy (Table 1).

The timing of biopsies varies in different centres. In our hospital, biopsies are performed weekly during the first 2 months, once every 2 weeks during the following 2 months, once per month during the following 8 months and, finally, one biopsy is performed every 6 months after the first year post-transplant [3].

For the management of transplant biopsies, we use a rapid-inclusion method, which enables us to make a diagnosis 4 h after receiving the specimen in our laboratory [3, 4].

2 Hyperacute Rejection

Vascularized organs (especially kidney and heart) are at risk of hyperacute rejection if the patient has pre-formed donor-specific alloantibodies [5]. A previous transplant, a blood transfusion or pregnancy may cause this humoral pre-sensitization (or alloimmunization). The primary target of the donor-specific antibodies is the vascular endothelium of the transplanted organ. The most common target antigens are human leukocyte antigen (HLA) class I and the ABO blood type system, but other less well-defined antigens have also been considered. HLA class-II antigens seem less relevant in hyperacute rejection because they are not strongly expressed on the vascular endothelium [6].

Antibodies mediating hyperacute rejection are almost always complement fixing and may be immunoglobulin G (IgG) or immunoglobulin M (IgM) type or both. Alloreactive T lymphocytes are occasionally involved. A critical step in alloantibody-mediated rejection is the initiation of the complement cascade, which leads to the release of various inflammatory mediators and the initiation of the coagulation and fibrinolytic systems [6].

Hyperacute rejection is an uncommon form of rejection that occurs immediately after transplantation. The morphological changes usually occur shortly after the patient is weaned from the bypass pump and the circulation is re-established. The histopathological changes of this condition are better documented in experimental animal models [7, 8]. Morphologically, the myocardium acquires a deep-red colour as a result of diffuse haemorrhage into the interstitium. Without emergent re-transplantation or mechanical support, the patient will not survive because of graft failure. If the patient survives for a few hours, diffuse interstitial oedema, neutrophilic infiltrates, fibrin thrombi and platelet aggregates may be found in small vessels, particularly in capillaries, arterioles and venules. Marginating neutrophil leukocytes are also present within the capillaries and venules.

3 Early Cardiac Allograft Failure

Early cardiac allograft failure (ECAF) seems best defined as a severe dysfunction of the cardiac graft in the absence of classic hyperacute rejection, acute cellular rejection or anticipated right ventricular dysfunction from pre-existing pulmonary hypertension. ECAF has been reported to occur in 4–25% of patients undergoing orthotopic heart transplantation. The highest risk of graft failure and overall mortality occurs during the first 30 days following transplantation [9].

In the first decades of the history of human transplantation, the main goal was to control acute rejection. This has been almost entirely achieved, thanks to the development of the EMB technique and to the consequent assessment of monitoring strategies. However, in the world series, half of the deaths still occur in the first post-operative month.

Ischaemic injury caused by prolonged ischaemic times, poor preservation, surgical problems, severe pulmonary hypertension or occult coronary artery disease in the donor may contribute to the development of this syndrome (Table 2). The role of reperfusion injury in the setting of orthotopic transplantation not well defined, but may contribute to poor graft function. In addition to ischaemic injury, humoral mechanisms have been implicated in ECAF [9].

Table 2. Causes of early allograft dysfunction

Primary graft failure:
Trauma
Metabolic
Catecholamine excess
Warm ischaemia
Poor preservation
Prolonged cold ischaemia
Hyperacute rejection
Technical problems
Fixed pulmonary hypertension
Acute cellular and/or humoral rejection

Biopsies from patients with haemodynamic compromise and graft dysfunction should also be carefully screened for histological evidence of vasculitis. Failure to find any evidence of cellular-mediated rejection may warrant fluorescent staining of the biopsy specimen with fluorescent antibodies against immunoglobulins and complement [9, 10]. It is uncommon for humoral rejection to occur in the absence of cellular rejection, except in multiparous women and previously transfused patients.

Morphologically, ECAF manifests with acute right ventricular dilation and pump failure. The pathological picture consists of interstitial myocardial oedema and haemorrhage, sometimes with fibrin thrombi in the arterioles. A frequent pathological feature in recipients who died of acute graft failure in the post-operative period is the presence of ischaemic necrosis, focal or diffuse [9–11]. Patchy ischaemic damage is usually attributable to the reperfusion injury, especially after a distant procurement of the donor heart, or the infusion of high doses of catecholamines.

4 Preservation Injury

Ischaemic injury may be caused by prolonged cold ischaemic time due to transportation of the donor heart, followed by reperfusion with damage to capillaries, or by failure to remove all air bubbles within the coronary circulation at transplantation. While procurement-related ischaemic injury is considered to be the most important factor, other mechanisms of myocyte injury are possible [12, 13]. Donor-associated injury may be secondary to central nervous system trauma and brain death, with resultant endogenous catecholamine-induced myocardial necrosis and/or exogenous administration of catecholamines [13, 14]. The dominant morphological pattern of catecholamine-induced injury is focal myocyte necrosis with contraction bands and histiocytic inflammation [14].

Preservation or ischaemia/reperfusion injury is seen early (within the first 3 weeks after transplantation) and is characterized by areas of ischaemic myocyte necrosis and/or variable areas of myocyte dropout (Fig. 1) [15]. These necrotic/dropout zones are usually associated with mild neutrophilic infiltration or no inflammation, in contrast with the mononuclear and eosinophilic infiltrate that commonly signals the onset of acute cellular rejection. In ischaemic injury, the degree of myocyte damage is not in proportion with the number of inflammatory cells in the interstitium (scattered polymorphonuclear leukocytes). Although infarcts occur, the ischaemic necrosis is usually subendocardial and focal, suggesting that the injury occurred while the heart was not filled with oxygenated blood. The affected myocytes show coagulative necrosis and a sharp border between polymorphous inflammation and nearby viable myocytes, shrinkage in size, hypereosinophilia with pyknotic nuclei, granular cytoplasm and, frequently, contraction bands. A trichrome stain will outline the ischaemic focus with a gray–blue colouration of the affected myocytes. Necrotic cells are either isolated or, more frequently, clustered into small groups. At a later stage, the necrotic myocardium is replaced by granulation tissue [15, 16]. It is reasonable to expect that severe, perioperative, ischaemic myocyte injury would markedly impair short-term function and sur-

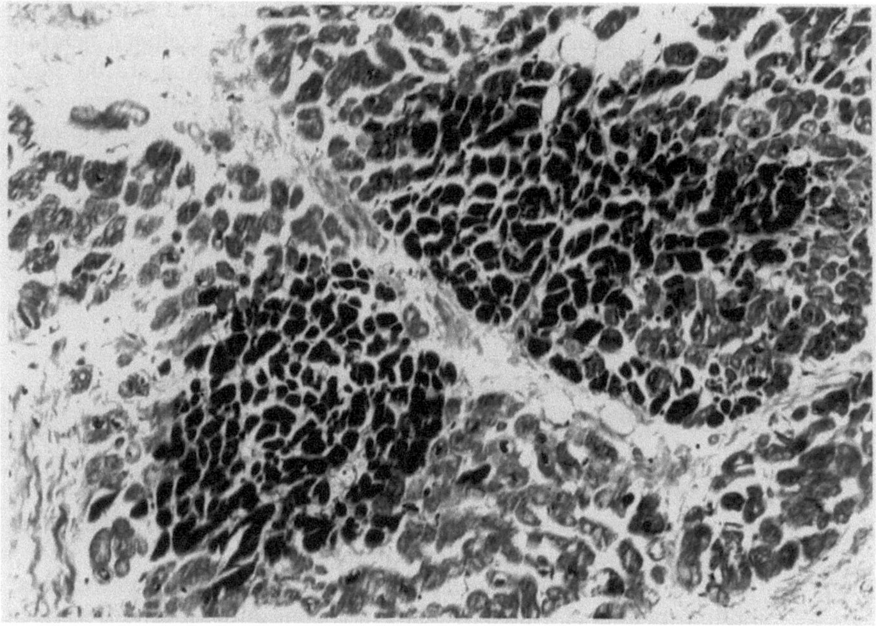

Fig. 1. Patchy reperfusion injury of the left ventricular myocardium. Focal circumscribed areas of coagulative necrosis without inflammatory infiltrate are evident. The recipient died of early cardiac allograft failure (ECAF) a few days after transplant (H & E, × 250)

vival after heart transplantation. Lesser degrees of ischaemic myocyte injury may contribute to late deterioration of function through myocardial fibrosis and possibly stimulate the development of graft atherosclerosis [13].

Ischaemic damage may also contribute to the development of graft atherosclerosis through endothelial injury, increased susceptibility to cytomegalovirus (CMV) infection and/or the promotion of rejection through release of donor alloantigens [13, 17].

5 Acute Cellular Rejection

The criteria for rejection of EMB biopsies are usually based on those initially proposed by Billingham, with modifications varying from institution to institution [18, 19]. Over the years, several different grading systems for acute rejection have been developed. In 1990, the Society for Heart and Lung Transplantation convened an international group of cardiac transplant pathologists in an attempt to develop a standardized grading system. The goal of this system was to provide a grading that was simple, easily taught and reproducible, as well as one that could be extrapolated to other grading systems. No matter which system is used, the following factors are assessed: (1) the nature, intensity and distribution of inflammatory cell infiltrates, (2) the presence or absence of oedema, and (3) the presence or absence of myocyte injury.

Cellular infiltrates with myocyte damage may also be observed in a variety of lesions that affect the myocardium and must, therefore, be excluded before a definite diagnosis of acute rejection is made [20]. The following are the more common pitfalls in the diagnosis of acute rejection: (1) the healing of ischaemic damage, (2) changes in old biopsy sites, (3) the Quilty-B effect, (4) infectious myocarditis, and (5) insufficient or inadequate tissue to rule out acute rejection.

Endomyocardial biopsies of acute cellular rejection reliably demonstrate myocardial inflammation and myocyte injury [21]. However, acute cellular rejection may also result in vascular inflammation [22]. Lymphocytes infiltrate vascular walls, but are not associated with endothelial disruption or vascular necrosis.

5.1 International Society for Heart and Lung Transplantation (ISHLT) Standardized Grading System

The ISHLT grading system [23] is described below and outlined in Table 3.

Grade 0. No evidence of acute rejection. Grade 0 represents normal myocardium without an inflammatory infiltrate or myocyte damage.

Grade 1. Grade 1A (focal, mild, acute rejection) represents a sparse perivascular infiltrate of large activated lymphocytes in one or two locations within a single biopsy fragment or in several biopsy pieces. Myocyte damage is not seen. Grade 1B (diffuse, mild, acute rejection) also represents a sparse, but more diffuse, infiltrate of lymphocytes extending into the interstitium between the myocytes and sometimes surrounding the myocytes, but without causing myocyte damage (Fig 2). It is not clear whether separation of 1A and 1B grades is clinically useful.

Grade 2. Grade 2 (focal, moderate, acute rejection) represents a single circumscribed infiltrate of lymphocytes with or without eosinophils and with focal myocyte damage (Fig. 3). This may occur by itself or may be accompanied by other forms of grade 1, but not higher grades. Early studies suggest that there is no difference in the outcome of grade 2, whether treated or not. The significance (ISHLT) of grade-2 cardiac allograft rejection has been questioned and the medical community is not in complete agreement as to its clinical management [24]. Fishbein et al. [25] have shown that most, if not all, cases of grade-2 cellular rejection can be shown to be Quilty-B lesions (see ppz) and are not associated with haemodynamic abnormalities and do not require augmented immunosuppression.

Grade 3. Grade 3 (moderate, acute rejection) consists of two types. Grade 3A represents multifocal lymphocytic infiltrates, with two or more foci causing myocyte damage or myocyte replacement (Fig. 4). Grade 3B indicates a more diffusely aggressive inflammatory infiltrate with myocyte necrosis within several pieces of the biopsy tissue (Fig. 5). It is sometimes designated as "borderline severe". The clinical usefulness of splitting biopsies into grade 3A or -3B has yet to be determined.

Grade 4. In grade 4 (severe, acute rejection), the inflammatory infiltrate is quite obvious and may become polymorphous with neutrophils and eosino-

Table 3. Standardized cardiac biopsy grading (ISHLT)

Nomenclature	Grade
No rejection	0
A = focal (perivascular or interstitial)	1
B = diffuse but sparse infiltrate	1
One focus only with aggressive infiltration and/or focal myocyte necrosis	2
A = multifocal aggressive infiltrates	3
B = diffuse inflammatory	3
Severe acute rejection	4
Resolving rejection	
Resolved rejection	

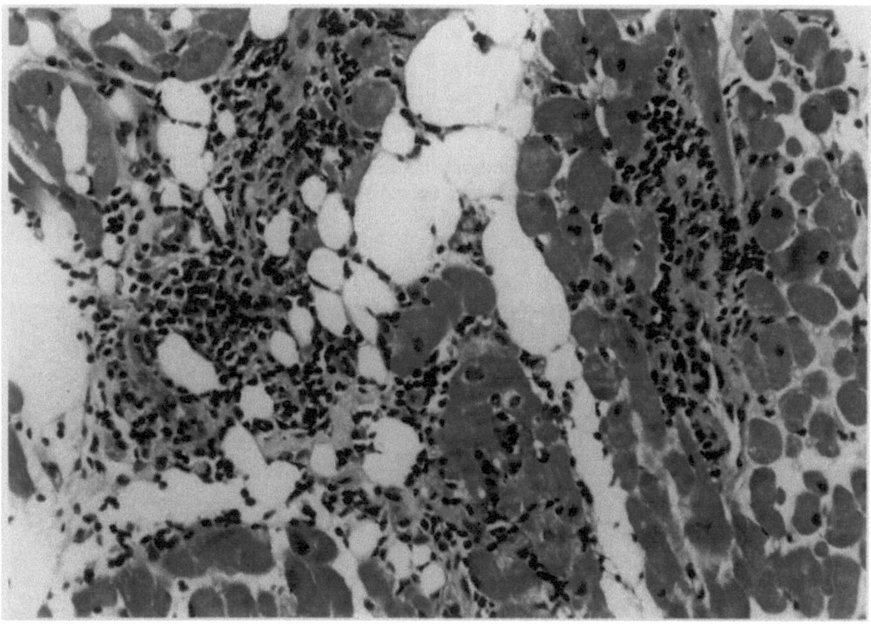

Fig. 2. Acute rejection grade IB (diffuse, mild, acute rejection). Sparse infiltration of lymphocytes extending into the interstitium between the myocytes but without necrosis (H & E, × 350)

phils. Myocyte necrosis is extensive (Fig 6). In some cases, an endothelialitis or vasculitis is seen; in these cases, interstitial haemorrhage is also present.

Resolving Rejection. This is designated by a lesser grade than in the previous biopsy showing rejection. There are often fibroblasts and iron-laden macrophages.

Resolved Rejection. This is grade ISHL 0 after a documented rejection episode and shows only mature scar, but no inflammatory infiltrate.

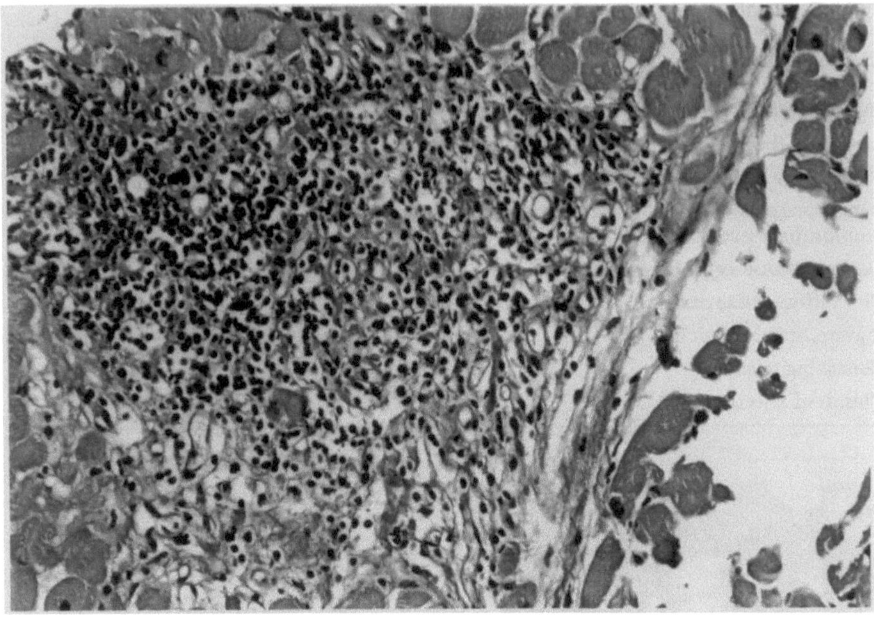

Fig. 3. Acute rejection grade II (focal, moderate acute rejection). A single circumscribed infiltrate of lymphocytes with focal cardiomyocyte necrosis (H & E, × 350)

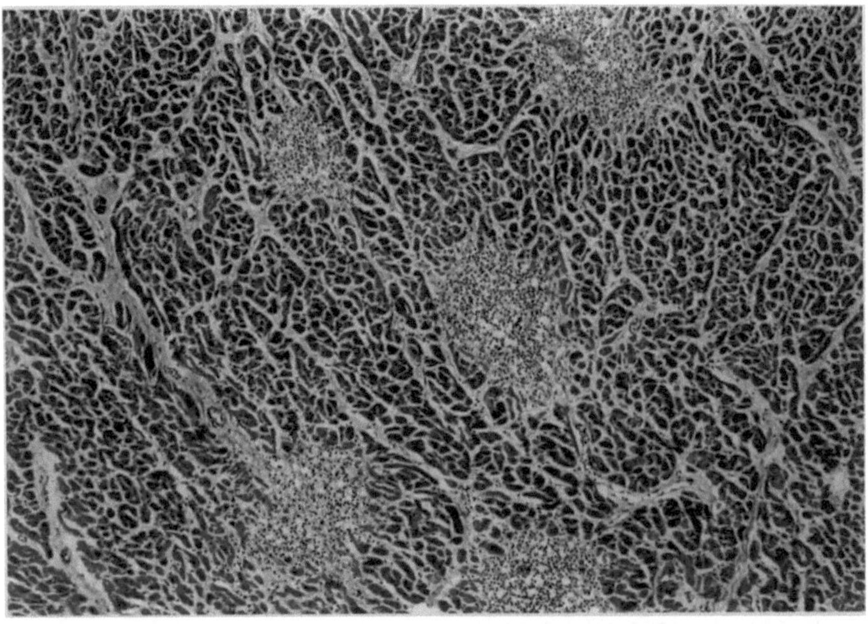

Fig. 4. Acute rejection grade IIIA. Multifocal lymphocytic infiltrates with myocyte replacement (H & E, × 100)

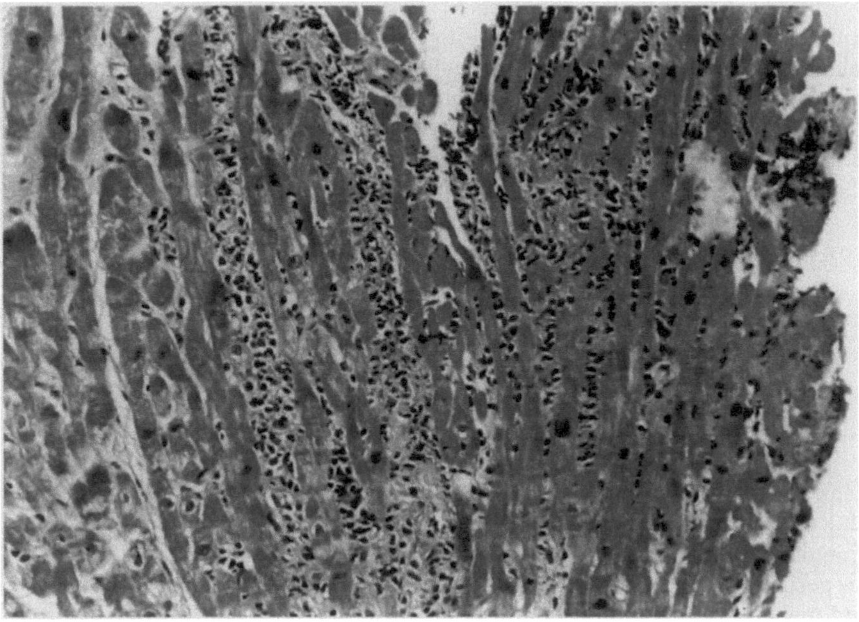

Fig. 5. Acute rejection grade IIIB. More extensive and aggressive inflammatory infiltrate with myocyte damage (H & E, × 250)

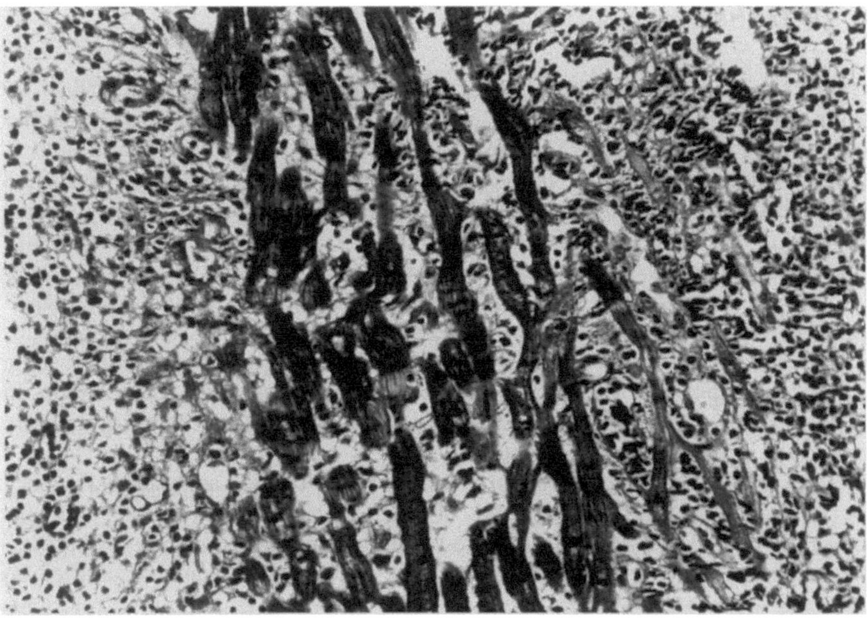

Fig. 6. Acute rejection grade IV, showing an extensive, mixed inflammatory infiltrate with damaged myocardium (H & E, × 350)

During rejection episodes, apoptosis of myocytes is one of the mechanisms of immune-mediated death, and it's investigation in tissue sections may represent a valuable tool for the diagnosis of myocyte damage [26]. Szabolcs et al. found that apoptosis of myocytes occurs during acute cardiac allograft rejection [27]. Apoptotic myocytes were localized adjacent to areas of inflammatory infiltration as well as in areas of a significant inflammatory infiltrate. Interaction of cytotoxic T lymphocytes with the Fas protein on macrophages and myocytes, as well as activation of cysteine proteases by granzyme-b, released from CD8-positive T cells, may also have caused myocyte apoptosis.

6 Previous Biopsy Sites

The average transplant recipient may undergo 15–20 biopsies during the first post-transplantation year. The bioptome tends to follow a similar path in any given patient. It is, therefore, common to take a biopsy specimen from a previous biopsy site [28]. Common features include partially organized fibrin masses, mildly inflamed granulation tissue, iron or ceroid-ladened macrophages and myocyte disarray at the periphery of the lesion. The cellular infiltrate associated with granulation tissue and myocyte disorientation in a healing biopsy site must be distinguished from acute rejection [3, 28, 29]. Lymphocytic infiltrate trapped in old biopsy sites may remain for years. Occasional foreign body granulomas can be seen [3].

7 Quilty Effect

Quilty lesions, also known as endocardial infiltrates, were first described by Billingham, who named them after the first patient showing this lesion [29]. Quilty lesions have been the subject of more than a dozen different studies and there is still no consensus as to their aetiology or significance. They have been associated with a direct drug interaction as these lesions were not observed with any significant degree of frequency before the addition of cyclosporin to the immunosuppressive regimen [30, 31]. It has also been suggested that Quilty lesions represent an unusual or unrecognized pattern of cellular or humoral "benign" form of rejection [32–34] or even an incipient Epstein-Barr virus (EBV)-associated lymphoproliferative process, the latter of which has been proven incorrect [35]; or a genetic predisposition associated with the *c-abl* oncogene. More recent studies in experimental animals suggest that they may be sites of antigen processing and low-grade immune stimulation [36].

The quilty effect (QE) typically manifest as flat or bulging aggregates of mononuclear cells composed predominantly of T cells ($CD4^+$ predominate over CD8 cells by a ratio of 2–3:1) and B lymphocytes with histiocytes, plasma cells and prominent microvasculature (Fig. 7) [3, 37]. Quilty lesions have been subclassified on the basis of whether they infiltrate the underlying myocardium. In type-A lesions, the border with the underlying myocardium is smooth. In type-B lesions, the mononuclear cells infiltrate the underlying myocardium, and this can cause

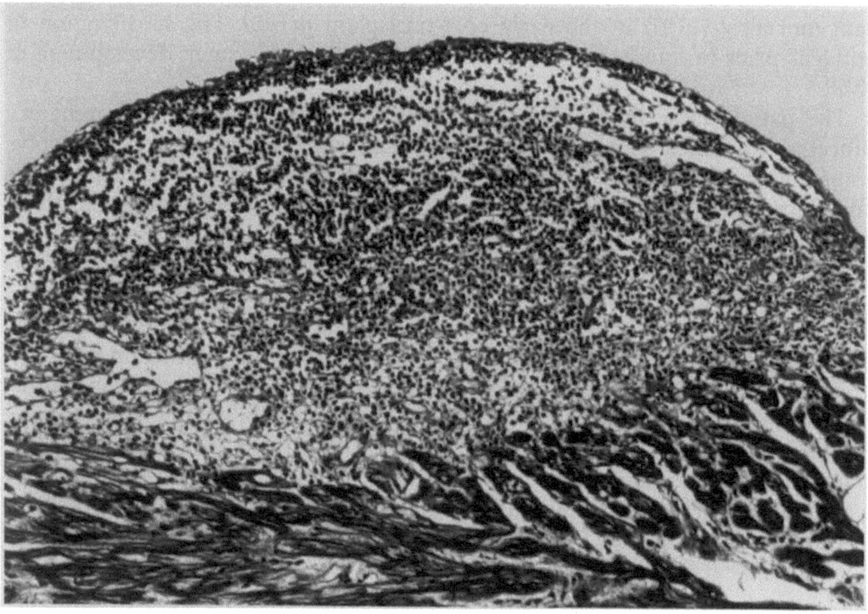

Fig. 7. Quilty effect. Collection of subendocardial lymphocytes without cardiomyocyte encroachment (type A) (H & E, × 200)

myocytolysis. On a practical level, quilty lesions are generally not considered in the grading of cardiac allograft rejection, but mentioned in the diagnosis as a separate finding. We have seen QE in biopsies taken between 7 days and 624 days after transplantation. It has also been noted that 47% of biopsies with QE also show different degrees of acute rejection (AR). Immunohistochemical studies show a similar type of lymphoid infiltrate in QE and AR; in both, most of the cells are T lymphocytes [37].

The nature of epicardial lymphoid infiltration in cardiac allografts and it's significance when observed in endomyocardial biopsies or autopsies is uncertain [38]. Epicardial lymphoid infiltrates occur with significant frequency after heart transplantation and can be associated with and mimic acute cellular rejection.

8 Humoral Rejection or Acute Vascular Rejection

Recently, investigators have observed clinical examples of patients with haemodynamic and echocardiographic evidence of graft dysfunction, but without the cardinal histological findings of cellular rejection. This is thought to represent "humoral" or antibody-mediated vascular rejection (AVR) [39]. Multiparous women are particularly at risk in the immediate post-transplant period, with onset of AVR most often in the first 2 weeks post-transplant. AVR is associated with clinically significant cardiac dysfunction in a majority of cases and carried significant mortality (17%) in the early post-transplant period. The sensitization to antigens prior to transplantation is important in the subsequent development of AVR.

The pathogenetic mechanisms responsible for these vascular inflammatory processes (in which an antigen-antibody complex process is implicated) involve complement activation, cytokine release, chemotaxis and activation of neutrophils and macrophages. All of these features suggest that delayed-type hypersensitivity may be implicated [39–41].

Endomyocardial biopsies taken at this time may show sparse acute inflammatory cells, severe interstitial oedema, haemorrhage, and endothelial swelling and injury of capillaries, venules and arterioles that may be subtle. Myocyte necrosis may be found in 60% of all AVR and in 100% of all clinically significant cases. On a larger scale, the biopsies may also show evidence of lymphocytic vasculitis. Damage to small vessels is often manifested by immunofluorescent evidence of the deposition of IgG or IgM, complement (C3 and/or C1q) and fibrinogen in a linear staining pattern around small vessels. Ig deposition may, however, be seen in ischaemia and other conditions and it is not specific for humoral rejection [39].

Hammond classified AVR into five grades on the basis of histological and immunocytochemical findings [39, 42]:

a) Negative or no evidence of AVR: all biopsies that show no light or immunofluorescent evidence of AVR.
b) Equivocal evidence of AVR: by light microscopy, biopsies show histological endothelial cell activation, and swelling or damage with or without associated oedema or haemorrhage. No inflammation or thrombosis is demonstrated. Immunofluorescence (IF) study shows microvascular accumulation of Ig or complement components, but not both. Similar equivocal findings can be seen in patients with systemic viral illness, especially those caused by CMV.
c) Mild AVR: light microscopic evidence of vasculitis may be demonstrated, which is often leukocytoclastic. By IF, the majority of cases show co-localization of Ig and complement in capillaries and venules. This group of patients consistently demonstrates alterations of the natural anticoagulant and fibrinolytic pathways.
d) Moderate AVR: in this category, vasculitis, arteriolitis, severe interstitial oedema and fibrin accumulation are seen. By IF, moderate microvascular rejection usually shows large accumulations of Ig and complement within capillaries and venules. In patients with moderate AVR, antithrombin III and tissue plasminogen activator (tPA) are usually completely lost. Clinically, this is often associated with haemodynamic compromise. Such biopsies may show piecemeal myocyte necrosis or subendocardial infarction.
e) Severe AVR: this is identical to severe acute cellular rejection and should be regarded as the end result of any severe rejection process. The EMB shows a diffuse mixed leukocytic infiltration including neutrophils and eosinophils. Myocyte necrosis, interstitial oedema and haemorrhage may be prominent. Vasculitis is obvious. IF often will have vascular deposits of Ig and complement, as well as interstitial and vascular accumulation of fibrin.

Mixed cellular and AVR may be seen in biopsy material and is independently graded, but simultaneously demonstrated. Hammond et al. [42] have found it prognostically useful to designate patients according to their predominant form of histological rejection. Patients with AVR have a significantly worse survival rate than patients with cellular or mixed-rejection patterns. It is interesting that although patients with mixed rejection have a survival rate similar to patients with cellular rejection, they have a fourfold higher risk of developing allograft coronary disease; patients with AVR have an eight- to ninefold greater risk [42, 43].

9 Infections

Cardiac allograft recipients are susceptible to a wide variety of bacterial, fungal, protozoal and viral infections [3]. These infections may involve the allograft itself, giving rise to infectious endocarditis, myocarditis and pericarditis. Infection is the second most common cause of death during the first year following heart transplantation (26%) [44]. Bacterial infections, particularly pneumonia, are the most common. The observation that the most common histological pattern of infection is necrotizing bronchopneumonia with abscess formation is not surprising, since the responsible pathogens are most often Pseudomonas and fungi. Two agents, *Toxoplasma gondii* and CMV, are of special interest because they are particularly likely to involve the allograft and may even be first detected on endomyocardial biopsy.

9.1 Toxoplasmosis

Toxoplasma gondii is an obligate intracellular parasite that is responsible for high mortality and morbidity rates [45]. Toxoplasmosis may develop as reactivation of a previous infection in seropositive patients, but toxoplasma seronegative recipients who receive a graft from a seropositive donor are at much higher risk of developing clinically significant toxoplasmosis [46]. Clinical disease may be limited to the heart, but in many cases there is dissemination and even lethal infection. Long-term pyrimethamine prophylaxis is recommended for seronegative recipients of hearts from seropositive donors. In our series, we found six cases of toxoplasmosis diagnosed by EMB in heart-transplant recipients. Despite treatment, one patient died with disseminated toxoplasmosis and pulmonary aspergillosis 10 days after diagnosis [47].

Our experience confirms other findings demonstrating that diagnosis of the disease can be difficult due to a lack of specific symptoms or signs and the lack of correlation of serological titres with clinical disease [47, 48]. The myocardium is known to be involved during the acute phase of toxoplasma infection and the EMB can be instrumental in the diagnostic process. Presently, molecular biology techniques, such as polymerase chain reaction (PCR) are joining the variety of procedures used in the diagnosis of toxoplasma infection.

Toxoplasmosis in the cardiac allograft mimics cellular rejection to a certain extent. In some cases, there are aggregates of inflammatory cells around individual necrotic myocytes. Lymphocytes and macrophages predominate, often with

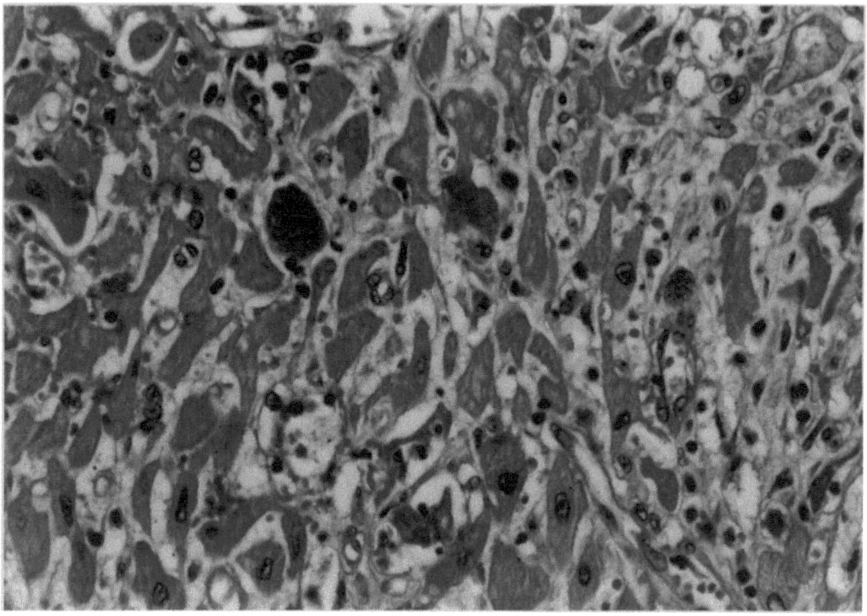

Fig. 8. Toxoplasma cysts in the sarcolemma of myocytes (H & E, × 350)

numerous eosinophils. However, eosinophils may be few in some cases of toxoplasmosis and prominent in some cases of cellular rejection. A careful search for the characteristic intramyocytic cysts is required (Fig. 8), even in areas lacking inflammation [3, 47], because cysts that have not ruptured, releasing the trophozoites, usually do not incite an inflammatory reaction

9.2 Cytomegalovirus

CMV infection is the most common and important viral infection in heart-transplant recipients, with incidences ranging from 73 % to 100 % [49]. Although many of these cases are asymptomatic – with diagnosis bsed on the shedding of virus or increase in serological titres – 25 % – 50 % of these patients will develop symptoms (fevers, malaise, leukopenia or invasive disease) that, in our experience, mandate treatment with ganciclovir. Disseminated CMV infection is associated with an increased frequency of graft rejection and bacterial and fungal infections. Viraemia and symptomatic CMV disease is more frequent in cardiac-transplant recipients than in other recipients. The risk of developing CMV infection is highest during weeks 4 – 12 post-operatively, and CMV occurs most frequently in recipients of organs from CMV-seropositive donors [50].

Even in patients with documented invasive or disseminated CMV infections, the virus is detected in only a fraction of EMB specimens. Involvement of the cardiac allograft is often less severe than that of other organs, such as the lungs, and endomyocardial biopsies are insensitive in detecting what is often focal disease [49, 50]. Nevertheless, necrotizing CMV myocarditis is occasionally recognized pathologically. Inflammation is usually relatively sparse and focal. A few lymphocytes, histiocytes and segmented neutrophils cluster around cells showing CMV inclusions or clumps of necrotic cellular fragments. A variety of cells, including endothelial cells, interstitial fibroblasts and myocytes may be infected. Cells showing diagnostic changes are enlarged and have a large central inclusion surrounded by a pale halo. Less distinct clumped cytoplasmic inclusions may also develop [51]. Immunohistochemistry for CMV early antigens or in situ hybridization for CMV nucleic acids and PCR may be required for positive identification in equivocal cases [51, 52].

We have seen three patients who died with disseminated CMV infection [3]. These three patients also had cerebral haemorrhage, pneumonia with positive culture for Legionella and disseminated lymphoma, respectively.

10 Malignancies in Heart Transplantation

The increased incidence of malignant tumours in heart transplant recipients has been well documented following the advent of immunosuppressive drugs. The incidence of novo malignant tumours varies from 1% to 16%, an incidence approximately 100 times greater than that of the matched general population [3, 53]. The high incidence of malignant disease among organ-transplant recipients is related to pharmacological suppression of the immune system, prolonged antigenic stimulation or oncogenic viral activation.

Distinctive tumours develop in different survival intervals. Non-Hodgkin's lymphomas are usually the earliest tumours to occur, even within the third postoperative month. Squamous-cell carcinomas (38%) and lymphomas (17%) are the most frequent neoplasms [53]. Kaposi's sarcoma is also associated with immunosuppression and EBV infection, but is very unusual in heart recipients [54].

10.1 Post-transplant Lymphoproliferative Disorders

Post-transplant lymphoproliferative disorders (PTLD or PT-LPD) encompass a range of lymphoid hyperplasias and neoplasms that occur in approximately 2%–5% of solid organ-transplant patients and about 1% of bone-marrow recipients. They have been described as "opportunistic neoplasms", which arise in part as a consequence of the immunosuppression required to prevent organ rejection [55, 56].

The characteristics of PTLD [57] in heart transplantion are:

- PTLD may arise in lymphoid tissues, such as tonsils or lymph nodes, but frequently arises in extranodal sites. The organ allograft itself is often involved by PTLD (Fig. 9).
- EBV is thought to be an important cofactor in the development of PTLD and is present in almost all cases (found with in situ hybridization [Fig. 10] and PCR techniques).

- Lymphomatous PTLDs are most often are of the non-Hodgkin's type.
- PTLDs most commonly arise from B-lymphocytes. Less commonly, tumours may arise from T cells or natural killer (NK) cells.
- The histological spectrum of PTLD ranges from hyperplastic lesions, some of which resemble mononucleosis, to atypical lymphoid lesions to lymphomas.
- Many cases regress, with reduction of immunosuppression and antiviral therapy.

Prior to 1981, lymphoid tumours in transplant patients were uniformly referred to as immunoblastic sarcomas. In that year, Frizzera and colleagues [58] from the University of Minnesota examined tumours from a small number of renal transplant recipients and observed several forms of lymphoproliferation that had not been previously described and applied the term "polymorphic" to emphasize the heterogeneity in size and shape of the tumour cells. The classification system of PTLD by Frizzera et al. includes reactive lymphoid hyperplasia, polymorphic diffuse B-cell hyperplasia, polymorphic B-cell lymphoma and immunoblastic sarcoma [58]. The authors stressed that the behavior of post-transplant lymphoproliferations could not be predicted reliably by pathological studies alone. In the series of Nalesnik and colleagues [55], Locker and Nalesnik [59] and Knowles et al. [60] have emphasized the monoclonality and molecular alterations of PTLDs.

The Society for Hematopathology Workshop, held in Duarte, California, in October 1995, led to discussions that expanded the spectrum of recognized post-transplant lymphoproliferations [61], including additional lesions such as plasmacytomas, T cell-rich B-cell lymphomas, and T-cell lymphomas under the umbrella term of PTLD.

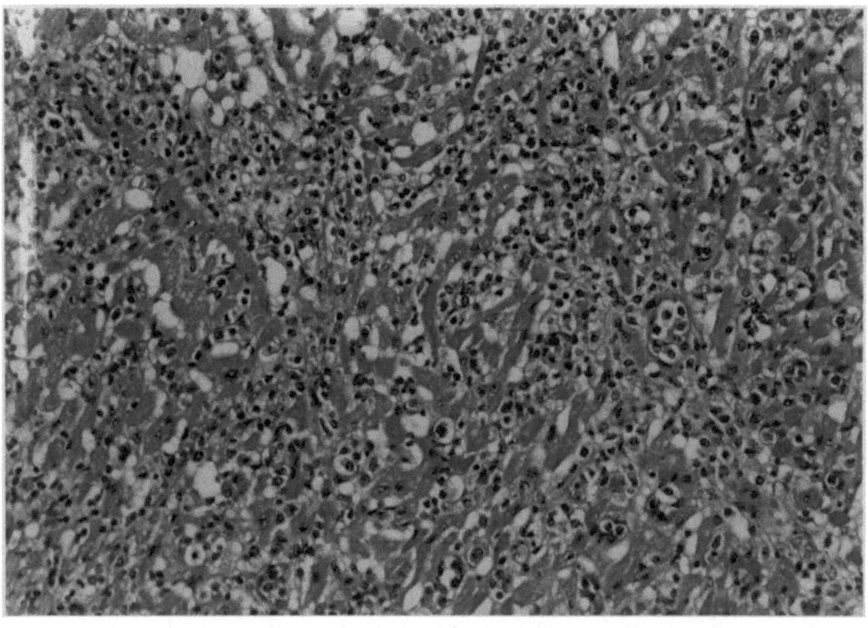

Fig. 9. Post-transplant lymphoproliferative disorders (PTLD) involving the heart. A polymorphous dense lymphoid infiltrate is seen between the myocytes (H & E, × 200)

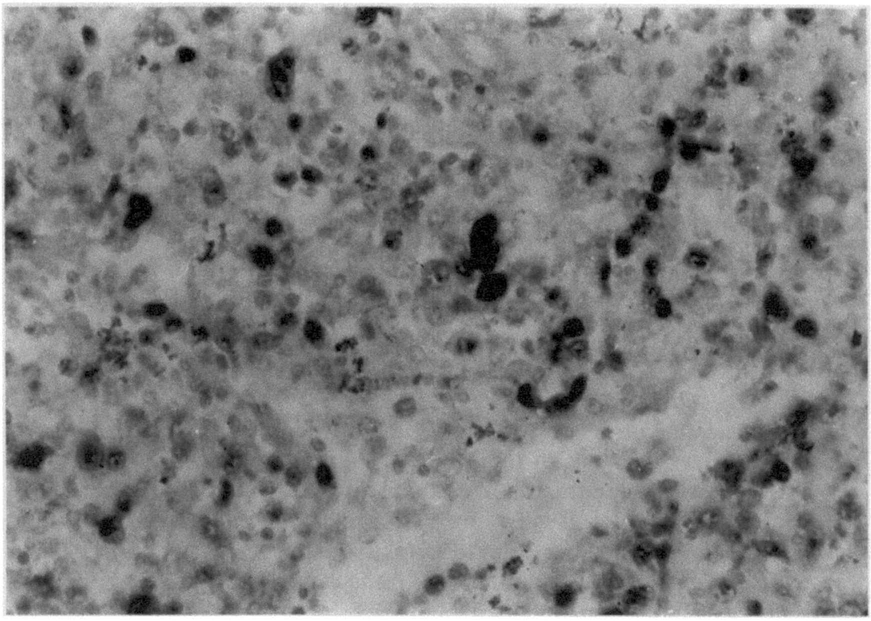

Fig. 10. In situ hybridization for Epstein-Barr virus (EBV) in a post-transplant lymphoproliferative disorders (PTLD) (×300)

10.2 Other Neoplasms

Skin and lung tumours are the most frequently occurring solid tumours in heart-transplant recipients. Skin tumours (except for Merkel's cell carcinoma and melanoma) usually have a benign course, whereas lung and other tumours developing in cardiac transplant recipients carry a poor prognosis [62]. Advanced disease stage at the time of diagnosis is responsible for the dismal outcome of recipients in whom solid tumours develop. Close post-operative tumour surveillance after cardiac transplantation is warranted.

The frequency of skin cancer in organ transplant recipients is high, up to 15%. In our series, between January 1984 and December 1993, at least one cutaneous neoplasm (squamous-cell carcinoma and/or basal-cell carcinoma) developed in 14 patients (15.2%). The basal-cell carcinoma to squamous-cell carcinoma ratio was 1:1.5. The skin cancer appeared an average of 31.5 months after transplantation; the average was 36 months for squamous-cell carcinoma and 25.3 months for basal-cell carcinoma. Cumulative risk rose from 4.3% at 1 year to 43.8% at 7 years after transplantation. The overall incidence of both types of skin cancer was 45.3 per 1000 post-transplant person–years, with an incidence of 25.8 for basal-cell carcinoma and 29.1 for squamous-cell carcinoma. Most skin cancers developed between 2 years and 3 years after transplantation [62]. Kaposi's sarcoma is fairly rare [54, 63]. Other tumours occur at a frequency similar to that of general population.

The association of EBV with smooth-muscle tumours was recently reported in the setting of acquired immunodeficiency syndrome (AIDS) and post-transplantation. In heart-transplant patients, two cases of EBV-associated smooth-muscle tumours arising in liver have been reported [64, 65]. The smooth-muscle tumour contained EBV within more than 95% of tumour cells when analyzed by means of Epstein-Barr (EBER1) in situ hybridization, was of strain type A when analyzed by means of Epstein-Barr nuclear antigen-2 (EBNA-2) PCR and contained an identical 30-bp deletion (amino acids 346–355) of the latent membrane protein-1 (LMP-1) oncogene when analyzed by PCR analysis.

11 Hypertrophy and Fibrosis

Cardiac transplantation is characterized by a sudden increase in ventricular loading, since the new heart has to restore cardiac output and normalize peripheral resistance, reversing interstitial fluid accumulation and pulmonary congestion in the recipient. Donors are usually young individuals of smaller cardiac weight; therefore, the transplanted heart has to adapt to the increased functional demand associated with these variables of the host. Transplanted hearts increase their size within weeks of transplantation. Heart biopsies show myocyte hypertrophy with large and hyperchromatic nuclei and giant mitochondria [66–68]. Beltrami et al. [69] suggest that the evolution of the transplanted heart involves the expression of a gene that is implicated in DNA replication. Proliferating cell nuclear antigen (PCNA)-labelled myocyte nuclei in the absence and presence of histological rejection was observed and no correlation existed between the severity of rejection and the percentage of myocytes and connective tissue cells stained by PCNA. This suggests that all these features support the idea that the proliferation of myocytes and non-muscle cells may be a component of ventricular remodelling after cardiac transplantation.

Myocardial fibrosis is a recognized post-transplantation event [3]. Studies have shown increased fibrosis in longer-surviving allografts. Fibrosis may be patchy or of a fine perimyocytic type. Previous rejection episodes or previous ischaemic episodes [70, 71] may cause patchy fibrosis. Li et al. [72] have described a relationship between the density of mast cells and their secreted products and the volume of fibrosis in post-transplant human hearts. The perimyocytic pattern of fibrosis is typical of cyclosporin-A effect [73–75]. Analysis of collagen volume fraction in biopsy material from allografts surviving more than 3 years showed that increased collagen deposition correlated with an increasing number of rejection episodes [76]. The authors postulated that damage from cellular rejection and more intensive cyclosporin-A therapy is responsible for increased fibrosis.

The other feature of fibrosis in long-term recipients involves the pericardium. Epicardial pathology is prevalent in longer-survival allografts. The pericardium may have become adhesive to the epicardium and thickened to the point at which it may cause constrictive pericarditis. Symmans et al. [66] have demonstrated that epicardial and pericardial fibrosis is often associated with nodular inflammatory infiltrates centred around neurovascular bundles. The significance of this is unclear, but a relationship between epicardial infiltrates and coronary artery pathology was found.

12 Denervation

After cardiac transplantation, the allograft remains denervated and is not under the usual control of the sympathetic autonomic nervous system [77]. More recently, there have been reports of some physiological evidence of re-innervation, although this has not yet been confirmed from the morphological standpoint. Some nerves can be seen in the transplanted heart, although they are greatly reduced in number from normal [78].

The conduction system in the transplanted heart is involved by the inflammatory infiltrate of the acute rejection. The small vessels to the sinus and atrioventricular (AV) node may be affected by graft coronary disease. The conduction system itself remains largely intact morphologically following cardiac transplantation, and there is little evidence of permanent structural damage [79]. It has been reported that there is about a 50% chance of surgical or procurement damage to both the recipient and the donor tissue around the conduction system [sino-atrial (SA) and AV nodes] [79, 80].

13 Calcification

Dystrophic, metastatic and idiopathic calcification of the myocardium has been reported sporadically [81]. Acute myocardial calcification of heart transplants was described by our group [82]. Several posterior cases have been reported [83]. Pulmonary hypertension, cardiac surgery and administration of steroids and calcium chloride, in addition to transitory acute renal failure may be the multi-causal factors that produce this myocardial calcification.

14 Chronic Rejection

The long-term survival of patients after orthotopic heart transplantation is limited by the development of obstructive coronary vascular lesions [chronic rejection (CR)] with resultant ischaemic myocardial injury. It similarly manifests in all vascularized solid-organ allografts as an obliterative arteriopathy or graft vascular disease, with interstitial fibrosis and atrophy of parenchymal elements, eventually resulting in allograft failure [84]. CR usually has an insidious onset, although abrupt arterial damage from a severe acute rejection can manifest similar arterial pathology. The principal histopathological finding in CR is concentric narrowing of the arterial lumen because of fibrointimal hyperplasia. Veins are much less frequently and less severely involved [84].

CR is seen as early as 3 months post-transplantation [84]. Most cases of clinically noticeable CR are demonstrable by coronary angiography 3–5 years post-transplantation. Accelerated graft arteriosclerosis occurs in approximately 35% of transplanted patients within 3 years after operation and in 50% within 5 years [84, 85]. Moreover, CR can develop and, unpredictably, become significantly obstructive within months of transplantation.

Table 4. Aetiologic aspects of Graft coronary disease (GCD)

Nonimmunological	Immunological and molecular
Age and gender	Acute rejection (humoral and cellular)
Corticosteroids	Transplantation antigens
Diabetes and hyperlipidaemia	Adhesion molecules
Estrogen	Growth factors and cytokines
Cytomegalovirus infection	New drugs

CR is thought to be due to direct immunological injury to the allogeneic arterial endothelium, which disrupts intimal homeostasis [86]. In turn, the injury is thought to trigger a cytokine- and growth factor-driven arterial repair response that results in luminal narrowing. Recent studies have outlined vascular endothelial-cell activation of various kinds (triggered by rejection and other process), including cytokines, growth factors, extracellular-matrix proteins, adhesion molecules and mediators such as interleukin-1, interleukin-2, platelet-derived growth factor, tumour growth factor-β and tumour necrosis factor-α, [85]. Non-immunological factors (Table 4) contributing to CR include hyperlipidaemia, diabetes mellitus, CMV infection and prolonged ischaemic time while harvesting the heart [85, 86–89]. CMV nucleic acids have been observed in the coronary arteries of allografted hearts, suggesting a possible role for the interaction of CMV with p53 in the development of accelerated graft arteriosclerosis in transplant recipients [90, 91]. CMV infection accelerates cardiac allograft arteriosclerosis, particularly in small intramyocardial arterioles mediated by inflammatory responses in the vascular wall and perivascular space [91]. By immunohistochemistry, Baas et al. [90] found that the smooth-muscle cells in these vessels did stain intensely for WAF-1. The expression of WAF1 further suggests that the WAF-1-mediated anti-proliferative signal is intact in CR vessels. Apoptosis and loss of functional vascular remodelling should be considered as important mediators of clinically relevant CR [92, 93].

Microscopic examination of CR shows varying degrees of concentric intimal proliferation (Fig. 11) and finally total occlusion of the lumen [3, 84, 85, 94, 95]. In the earlier stages, the intimal proliferation presents as a homogeneous appearance with varying amounts of transformed smooth-muscle cells, lymphoblasts, fibroblasts, and macrophages. Later, a layer of foamy macrophages lined up along the internal elastic membrane and then spreads out into the intima. An increase in extracellular matrix was also observed [96, 97]. Within 10 years post-transplantation, the intima may form typical plaque-like changes with atheroma and occasionally with calcium and cholesterol clefts. The endothelium usually remains intact and may have a lymphocyte component often referred to as endothelialitis. The internal elastic, unlike that in atherosclerosis, remains almost intact, except for very small breaks, through which the smooth-muscle cells apparently emigrate to the intima. In this process, the media remains intact and is apparently unaffected, except for the fact that, as the lumen of the artery becomes obliterated, the width of the media appears to become thinner. This type of intimal proliferation often affects large branches and penetrates intramuscular small vessels all the way to the endocardium [98, 99]. For this reason, small pyramidal infarcts are often found. Large full-wall thickness myocardial infarcts are also present in advanced CR and,

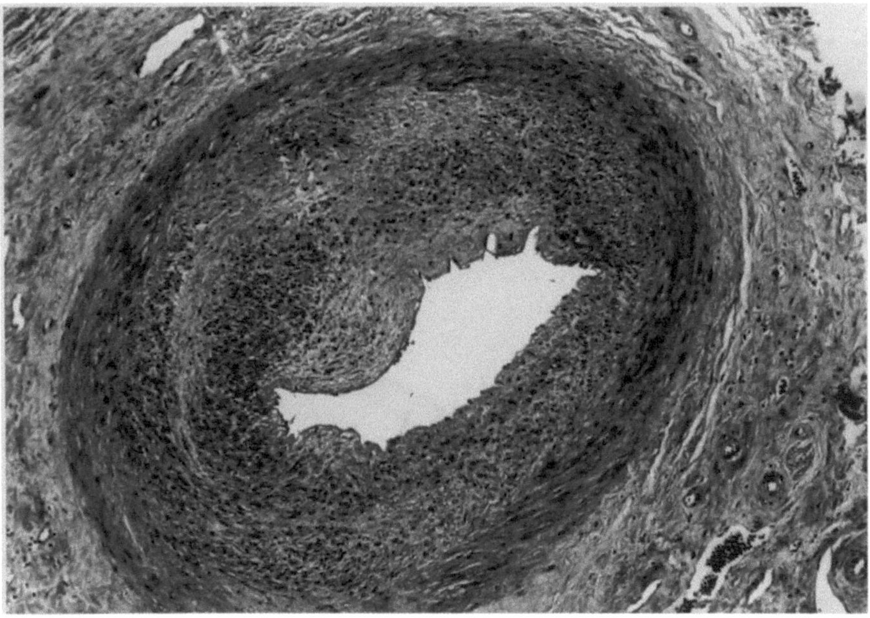

Fig. 11. Explanted heart showing an epicardial coronary artery with chronic rejection (concentric intimal proliferation) (H & E, × 50)

in most cases, these are clinically silent and unsuspected (because transplanted hearts are denervated). CR involving the venous system of the heart is very rare and mild.

A lack of correlation between microvascular and epicardial vessel disease suggests discordant manifestation and progression of CR [98]. Heart biopsies in graft atherosclerosis show myointimal proliferation, thickening and folding of the wall of pre-capillary arteries, myocyte hypertrophy and interstitial and perivascular fibrosis (Fig. 12) [100–102]. Type-III and type-IV collagen, laminin and fibronectin are increased in areas of interstitial and perivascular fibrosis, as well as in the vessel's wall. Fibronectin accumulation is more evident in the subendothelium and inner media of affected vessels [96].

Although the arterial changes in CR have been extensively reported, less attention has focused on the significance of the myocardial pathology resulting from the perfusion defects caused by CR. These changes include subendocardial myocyte vacuolization, indicative of sublethal ischaemic injury, and coagulative myocyte necrosis, acute and healing, indicative of infarction [101, 102]

Although accelerated arteriosclerosis is the most critical and most widely recognized complication in the allograft heart, a new subtype of coronary arteriopathy has been reported recently [103]. This complication, termed dilated angiopathy, was observed in about 7% of heart transplants undergoing coronary angiography 1 year or more after transplantation. The aneurysmal dilatation may involve any of the proximal epicardial coronary arteries and is not accompanied by either diffuse or discrete obliterations of the vascular lumen. Because histological specimens

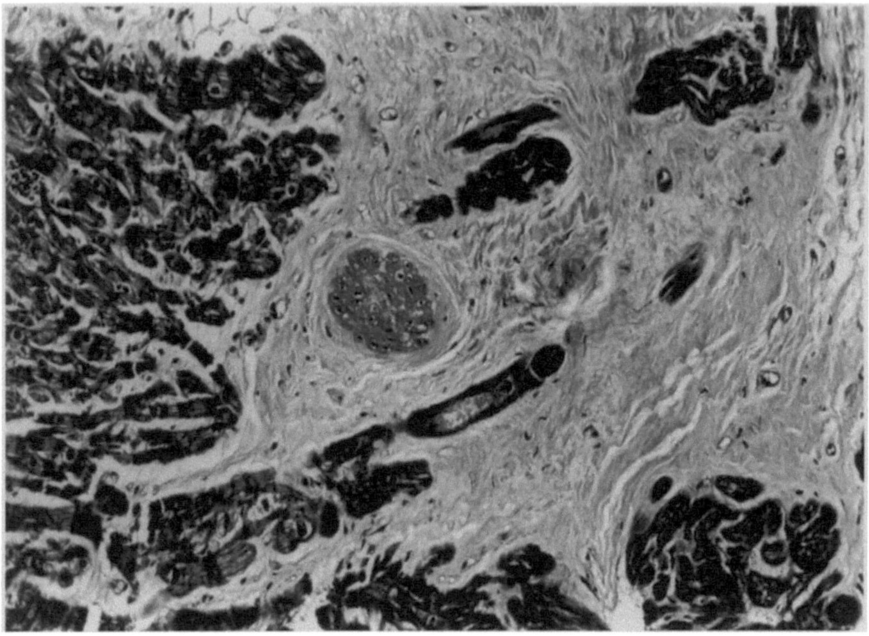

Fig. 12. Heart biopsy in chronic rejection shows myointimal proliferation of a small artery, myocyte hypertrophy, and interstitial and perivascular fibrosis (H & E, × 250)

were not available, it is unclear whether the underlying morphological alteration in dilated angiopathy is that of active or healed arteritis or some other unspecified medial degenerative process.

Although CR is similar to atherosclerosis seen in the general population, there are also distinct differences. A comparison of the two is shown in Table 5.

Balloon angioplasty has a limited role in the treatment of focal lesions. Experiences with coronary stenting, coronary-artery bypass grafting and trans-myocardial laser revascularization are limited [104]. Re-transplantation has a worse prognosis than initial transplantation. In addition, a recent study has shown that rapid or fulminant development of CR within 1 year foretells a poor prognosis as a result of ischaemic cardiac events [105]. Strategies for blocking expression of adhesion molecules may help prevent chronic rejection in clinical transplantation. 3-Hydroxy-3-methylglutaryl coenzyme-A reductase inhibitors and anti-proliferative drugs may slow progression of CR by a variety of effects [85].

15 Causes of Death

Despite the wealth of experience over the last 25 years, the major causes of death in cardiac allograft remain the same; acute cardiac rejection or infections in the early post-operative years or CR disease in the later years (Table 6).

Table 5. Comparison of Graft vascular disease (GVD) and atherosclerosis in the general population

Histopathological finding	Graft vascular disease	Atherosclerosis
Epicardial arteries	Involved	Preferentially involved
Penetrating intramyocardial arteries	Involved	Not involved
Endothelium	Often intact, but hypertrophied	Usually intact, hypertrophy not as obvious as GVD
Myointimal proliferation and luminal narrowing	Yes, concentric	Yes, eccentric
Intimal lipid and cholesterol deposits	Uncommon	Common
Intimal inflammation	Variable	Variable
Elastic lamina	Focally disrupted	Focally disrupted
Media	Thinned in late stage	Thinned in late stage
Medial inflammation	Variable	Variable
Adventitial inflammation	Common	Variable
Calcium deposition	Absent	Frequently present
Rate of development	Months	Years

Table 6. Causes of death in cardiac recipients. The prime causes of death are subdivided by survival intervals

Perioperative death (≤1 month)	Early death (>1 month, ≤3 months)	Intermediate death (>3 months, ≤2 years)	Late death (>2 years)
Graft failure: pump failure	Graft failure: pump failure	Graft arteriopathy	Graft arteriopathy
Multi-organ failure	Acute rejection	Infections: viral, bacterial, fungal	Tumours: lymphoma, carcinomas
Pulmonary hypertension	Infections: viral, fungal	Tumours: Pulmonary and skin carcinomas, lymphoma	
Postoperative complications	Tumours: lymphoma		
Bacterial infections			

It is well known that ischaemic [106], valvular [106] or congenital heart diseases [107] and myocarditis [108] adversely affect transplant outcome. Patients transplanted for idiopathic dilated cardiomyopathy had a significantly better outcome [109].

The main obstacle to long-term survival is now represented by graft vasculopathy. However, in the series, nearly three-fourths of the deceased recipients died within the first 6 post-operative months, and half of deaths still occur in the first post-operative month [109–111]. Deaths are mainly attributable to acute graft failure, post-operative complications, non-cardiac emergencies and bacterial infections in the first post-operative month, and to acute rejection and saprophytic infections in the subsequent months. However, myocardial dysfunction and ischaemic necrosis secondary to graft vasculopathy usually occur before the end of the sixth post-operative month.

Fatal infections are a frequent occurrence in short-term survivors [110], but the distribution of the agents varies with different periods [112]; although bacterial infections are more common in the first 3 post-operative months, saprophytes predominate later. The relatively high prevalence of fatal mycotic infections is consistent with the low mortality from acute rejection and warns against the risk of excessive immunosuppressive treatment of recipients. Viral infections seem to play an initiating role because they are slightly more common as the primary than the immediate cause of death.

References

1. Caves PK, Stinson EB, Billingham ME, et al. (1974) Serial transvenous biopsy of the transplanted human heart. Improved management of acute rejection episodes. Lancet 1: 821–826
2. Fowles RE (1978) Techniques for right and left ventricular endomyocardial biopsy. Am J Cardiol 41: 887–892
3. Pardo Mindán FJ, Lozano MD, Contreras-Mejuto F, de Alava E (1992) Pathology of heart transplant through endomyocardial biopsy. Semin Diagn Pathol 9: 238–248
4. Pardo Mindán J (1987) Utilidad de la inclusión rápida de la biopsia renal por punción en el proceso de toma de decisiones en el trasplante renal. In: Caralps A, Griñó JM, Brulles A, et al. (eds) Trasplante de Organos y Tejidos. DOYMA, Barcelona, Spain, pp 133–152
5. Weil R, Clarke DR, Iwaki Y, Porter KA, Koep LJ, Paton BC, Terasaki PI, Starzl TE (1981) Hyperacute rejection of a transplanted human heart. Transplantation 32: 71–72
6. Trento A, Hardesty RL, Griffith BP, Zerbe T, Kormos RL, Bahnson HT (1988) Role of the antibody to vascular endothelial cells in hyperacute rejection in patients undergoing cardiac transplantation. J Thorac Cardiovasc Surg 95: 37
7. Forbes RDC, Kuramochi T, Guttman RD, Klassen J, Knaack J (1975) A controlled sequential morphologic study of hyperacute cardiac allograft rejection in the rat. Lab Invest 33: 280–288
8. Fryer JP, Leventhal JR, Dalmasso AP, Chen S, Simone PA, Goswitz JJ, Reinsmoen NL, Matas AJ (1995) Beyond hyperacute rejection. Accelerated rejection in a discordant xenograft model by adoptive transfer of specific cell subsets. Transplantation 59: 171–176
9. Hauptman PJ, Aranki S, Mudge GH, Couper GS, Loh E (1994) Early cardiac allograft failure after orthotopic heart transplantation. Am Heart J 127: 179–186
10. Gallo P, Agozzino L, Arbustini E, Bartoloni G, Baroldi G, Bonacina E, Bosman C, Catani G, di Gioia C, Motta T, Pucci A, Rocco M, Thiene G (1994) Immediate causes of death in short-term surviving heart transplant recipients. Cardiovasc Pathol 3: 173–181
11. Fernandez AL, Herreros JM, Llorens R, Martinez A, Panizo A, Manito N (1996) Primary graft failure after heart transplantation. Successful recovery with pneumatic biventricular assistance. Int J Artif Organs 19: 307–310
12. Ballester M, Obrador D, Abadal L, Cladellas M, Bordes R, Manito N, Pons-Llado G, Padro JM, Aris A, Caralps-Riera JM (1989) Dopamine treatment of locally procured donor hearts: relevance on postoperative cardiac histology and function. Int J Cardiol 22: 37–42
13. Knight RJ, Dikman S, Liu H, Martinelli GP (1997) Cold ischemic injury accelerates the progression to chronic rejection in a rat cardiac allograft model. Transplantation 64: 1102–1107

14. Baroldi G, Di Pasquale G, Silver MD, Pinelli G, Lusa AM, Fineschi V (1997) Type and extent of myocardial injury related to brain damage and its significance in heart transplantation: a morphometric study. J Heart Lung Transplant 16: 994–1000
15. Fyfe B, Loh E, Wionters GL, Couper GS, Kartashov AI, Schoen FJ (1996) Heart transplantation-associated perioperative ischemic myocardial injury. Morphological features and clinical significance. Circulation 93: 1133–1140
16. Garcia-Poblete E, Fernandez H, Alvarez L, Torralba A, Escudero C (1997) Structural and ultrastructural study of the myocardium after 24-hour preservation in University of Wisconsin solution. Histol Histopathol 12: 375–382
17. Gaudin PB, Rayburn BK, Hutchins GM, Kasper EK, Baugham KL, Goodman SN, Lecks LE, Baumgartner WA, Hruban RH (1994) Peritransplant injury to the myocardium associated with the development of accelerated arteriosclerosis in heart transplant recipients. Am J Surg Pathol 18: 338–346
18. Billingham ME, Cary NRB, Hammond ME, et al. (1990) A working formulation for the standarization of nomenclature in the diagnosis of heart and lung rejection: heart rejection study group. J Heart Transplant 9: 587–593
19. Lozano MD, Pardo-Mindán FJ, Herreros J (1989) Value of Hannover's classification of acute cardiac allograft rejection. Clin Transplant 3: 336–340
20. Billingham ME (1981) Diagnosis of cardiac rejection by endomyocardial biopsy. J Heart Transplant 1: 25–30
21. Winters GL (1997) The challenge of endomyocardial biopsy interpretation in assessing cardiac allograft rejection. Curr Opin Cardiol 12: 146–152
22. Herskowitz A, Soule LM, Ueda K, et al. (1987) Arteriolar vasculitis on endomyocardial biopsy: a histologic predictor of poor outcome in cyclosporine treated heart transplant recipients. J Heart Transplant 6: 127–136
23. McAllister HA Jr. (1990) Histologic grading of cardiac allograft rejection: a quantitative approach. J Heart Transplant 9: 277–282
24. Brunner-La Rocca HP, Sutsch G, Schneider J, Follath F, Kiowski W (1996) Natural course of moderate cardiac allograft rejection (International Society for Heart Transplantation grade 2) early and late after transplantation. Circulation 94: 1334–1338
25. Fishbein MC, Bell G, Lones MA, Czer LS, Miller JM, Harasty D, Trento A (1994) Grade 2 cellular heart rejection: does it exist? J Heart Lung Transplant 13: 1051–1057
26. Laguens RP, Meckert PM, Martino JS, Perrone S, Favaloro RJ (1996) Identification of programmed cell death (apoptosis) in situ by means of specific labeling of nuclear DNA fragments in heart biopsy samples during acute rejection episodes. Heart Lung Transplant 15: 911–918
27. Szabolcs M, Michler RE, Yang X, Aji W, Roy D, Athan E, Sciacca RR, Minanov OP, Cannon PJ (1996) Apoptosis of cardiac myocytes during cardiac allograft rejection. Relation to induction of nitric oxide synthase. Circulation 94: 1665–1673
28. Winters GL (1991) The pathology of heart allograft rejection. Arch Pathol Lab Med 115: 266–272
29. Billingham ME (1990) Cardiac transplantation. In: Sale GE (ed) The pathology of organ transplantation. Butterworths, Boston, pp 133–152
30. Suit PF, Kottke-Marchant K, Ratliff NB, et al. (1989) Comparison of whole-blood cyclosporine levels and the frequency of endomyocardial lymphocytic infiltrates (the quilty lesion) in cardiac transplantation. Transplantation 48: 618–621
31. Forbes RDC, Rowan RA, Billingham ME (1990) Endocardial infiltrates in human heart transplants: a serial biopsy analysis comparing four immunosuppression protocols. Hum Pathol 21: 850–855
32. Joshi A, Masek MA, Brown BW Jr, et al. (1995) "Quilty" revisted: a 10-year perspective. Hum Pathol 26: 547–557
33. Kemnitz J, Cremer J, Schaefers H, et al. (1991) Some aspects of changed histopathologic appearance of acute rejection in cardiac allografts after prophylactic application of OKT3. J Heart Lung Transplant 10: 366–372
34. Pardo-Mindán FJ, Lozano MD (1991) "Quilty effect" in heart transplantation: is it related to acute rejection? J Heart Lung Transplant 10: 937–941

35. Nakhleh RE, Copenhaver CM, Werdin K, et al. (1991) Lack of evidence for involvement of Epstein-Barr virus in the development of the "quilty" lesion of transplanted hearts: an in situ hybridization study. J Heart Lung Transplant 10: 504–507
36. Demetris AJ, Murase N, Ye Q, et al. (1997) Analysis of chronic rejection and obliterative arteriopathy. Possible contributions of donor antigen presenting cells and lymphatic disruption. Am J Pathol 150: 563–578
37. Lozano MD, Pardo-Mindán FJ (1994) Significance of endocardial infiltrates (Quilty effect) in the transplanted heart. J Heart Lung Transplant 13: 733–734
38. Luthringer DJ, Yamashita JT, Czer LS, Trento A, Fishbein MC (1995) Nature and significance of epicardial lymphoid infiltrates in cardiac allografts. J Heart Lung Transplant 14: 537–543
39. Hammond EH, Hansen JH, Spencer LS, et al. (1993) Vascular rejection in cardiac transplantation: histologic, immunopathologic, and ultrastructural features. Cardiovasc Pathol 2: 21–34
40. Caple JF, McMahon JT, Myles JL, Hook S, Ratliff NB (1995) Acute vascular (humoral) rejection in non-OKT3-treated cardiac transplants. Cardiovasc Pathol 4: 13–18
41. Hauptman PJ, Aranki S, Mudge GH Jr, Couper GS, Loh E (1994) Early cardiac allograft failure after orthotopic heart transplantation. Am Heart J 127: 179–186
42. Hammond EH, Yowell RL, Price GD, et al. (1992) Vascular rejection and its relationship to allograft coronary disease. J Heart Transplant 11: S111
43. Ensley RD, Hammond EH, Renlund DG, et al. (1991) Clinical manifestations of vascular rejection in cardiac transplantation. Transplant Proc 23: 1130
44. Graham AR (1992) Autopsy findings in cardiac transplant patients. A 10-year experience. Am J Clin Pathol 97: 369–375
45. Holliman RE, Johnson JD, Adams S, et al. (1991) Toxoplasmosis and heart transplantation. J Heart Lung Transplant 10: 608
46. Dressler FA, Javier JJ, Salinas-Madrigal L, Milligan TW, McBride LR, Labovitz AJ, Miller LW (1996) Myocardial toxoplasmosis complicating cardiac transplant. Cardiovasc Pathol 5: 101–104
47. Lozano MD, Pardo-Mindán FJ (1996) Value of the endomyocardial biopsy in the diagnosis of topxoplasmosis after heart transplantation. Cardiovasc Pathol 5: 55–56
48. Arnold SJ, Kinney MC, McCormick MS, Dummer S, Scott MA (1997) Disseminated toxoplasmosis. Unusual presentations in the immunocompromised host. Arch Pathol Lab Med 121: 869–873
49. Millett R, Tomita T, Marshall HE, et al. (1991) Cytomegalovirus endomyocarditis in a transplanted heart. Arch Pathol Lab Med 115: 511–515
50. Dummer JS, White LT, Ho M, et al. (1985) Morbidity of cytomegalovirus infection in recipients of heart or heart-lung transplants who received cyclosporine. J Infect Dis 152: 1182–1191
51. Pucci A, Ghisetti V, Donegani E, Barbui A, David E, Fortunato M, Papandrea C, Pansini S, Zattera G, di Summa M, et al. (1994) Histologic and molecular diagnosis of myocardial human cytomegalovirus infection after heart transplantation. J Heart Lung Transplant 13: 1072–1080
52. Weiss LM, Movahed LA, Berry GJ, Billingham ME (1990) In situ hybridization studies for viral nucleic acids in heart and lung allograft biopsies. Am J Clin Pathol 93: 675–679
53. Pham SM, Kormos RL, Landreneau RJ, Kawai A, Gonzalez-Cancel I, Hardesty RL, Hattler BG, Griffith BP (1995) Solid tumors after heart transplantation: lethality of lung cancer. Ann Thorac Surg 60: 1623–1626
54. Aebischer MC, Zala LB, Braathen LR (1997) Kaposi's sarcoma as manifestation of immunosuppression in organ transplant recipients. Dermatology 195: 91–92
55. Nalesnik MA, Jaffe R, Starzl TE, Demetris AJ, Porter K, Burnham JA, Makowka L, Ho M, Locker J (1988) The pathology of posttransplant lymphoproliferative disorders occurring in the setting of cyclosporine A-Prednisone immunosuppression. Am J Pathol 133: 173–192
56. Lozano MD, Pardo-Mindán FJ, Contreras F, Flórez I, Herreros J (1991) Value of endomyocardial biopsy in the diagnose of lymphoma after heart transplantation. A case report. Clin Transplant 5: 33–35

57. Kowal-Vern A, Swinnen L, Pyle J, Radvany R, Dizikes G, Michalov M, Molnar Z (1996) Characterization of postcardiac transplant lymphomas. Histology, immunophenotyping, immunohistochemistry, and gene rearrangement. Arch Pathol Lab Med 120: 41–48
58. Frizzera G, Hanto DW, Gajl-Peczalska KJ, Rosai J, McKenna RW, Sibley RK, Holahan KP, Lindquist LL (1981) Polymorphic diffuse B-cell hyperplasias and lymphomas in renal transplant recipients. Cancer Res 41: 4262–4279
59. Locker J, Nalesnik M (1989) Molecular genetic analysis of lymphoid tumors arising after organ transplantation. Am J Pathol 135: 977–987
60. Knowles DM, Cesarman E, Chadburn A, Frizzera G, Chen J, Rose EA, Michler RE (1995) Correlative morphologic and molecular genetic analysis demonstrates three distinct categories of posttransplant lymphoproliferative disorders. Blood 85: 552–565
61. Harris NL, Ferry JA, Swerdlow SH (1997) Posttransplant lymphoproliferative disorders: summary of Society for Hematopathology Workshop. Semin Diagn Pathol 14: 8–14
62. España A, Redondo P, Fernandez AL, Zabala M, Herreros J, Llorens R, Quintanilla E (1995) Skin cancer in heart transplant recipients. J Am Acad Dermatol 32: 458–465
63. Lopez-Rubio F, Anguita M, Arizon JM, Lopez-Beltran A, Mesa D, Lopez-Granados A, Valles F, Concha MJ (1994) Visceral Kaposi's sarcoma without mucocutaneous involvement in a heart transplant recipient. Heart Lung Transplant 13: 913–915
64. Kingma DW, Shad A, Tsokos M, Fest T, Otsuki T, Frekko K, Werner E, Werner A Magrath I, Raffeld M, Jaffe ES (1996) Epstein-Barr virus (EBV)-associated smooth-muscle tumor arising in a post-transplant patient treated successfully for two PT-EBV-associated large-cell lymphomas. Case report. Am J Surg Pathol 20: 1511–1519
65. Davidoff AM, Hebra A, Clark BJ 3rd, Tomaszewski JE, Montone KT, Ruchelli E, Lau HT (1996) Epstein-Barr virus-associated hepatic smooth muscle neoplasm in a cardiac transplant recipient. Transplantation 61: 515–517
66. Symmans WF, Nielsen H, Dell R, Rose E, Marboe CC (1994) Cardiac allograft pathology: a clinicopathologic correlation. Cardiovasc Pathol 3: 249–256
67. Imakita M, Tazelaar HD, Rowan RA, Masek MA, Billingham ME (1987) Myocyte hypertrophy in the transplanted heart. A morphometric analysis. Transplantation 43: 839–842
68. Rowan RA, Billingham ME (1990) Pathologic changes in the long-term transplanted heart. Hum Pathol 21: 767–772
69. Beltrami CA, Di Loreto C, Finato N, Rocco M, Artico D, Cigola E, Gambert SR, Olivetti G, Kajstura J, Anversa P (1997) Proliferating cell nuclear antigen (PCNA), DNA synthesis and mitosis in myocytes following cardiac transplantation. J Mol Cell Cardiol 29: 2789–2802
70. Pickering JG, Boughner DR (1990) Fibrosis in the heart and its association with ischemic time. Circulation 81: 949–958
71. Chomette G, Auriol M, Delcourt A, Karkouche B, Cabrol A, Cabrol C (1985) Human cardiac transplants: diagnosis of rejection by endomyocardial biopsy: causes of death. Virchows Arch 407: 295–307
72. Li QY, Raza-Ahmad A, MacAulay MA, Lalonde LD, Rowden G, Trethewey E, et al. (1992) The relationship of mast cells and their secreted products to the volume of fibrosis in post-transplant hearts. Transplantation 53: 1047–1051
73. Karch SB, Billingham ME (1985) Cyclosporine induced myocardial fibrosis: a uniquely controlled case report. J Heart Transplant 4: 210–212
74. Stovin PGI, English TAH (1985) Effects of cyclosporine on the transplanted human heart. J Heart Transplant 6: 180–185
75. Myles JL, Ratliff NB, McMahon JT, et al. (1988) Cyclosporin-associated microfibrils in cardiac transplant patients. Am J Cardiovasc Pathol 127–132
76. Tazelaar HD, Gay RE, Rowan RA, Billingham ME, Gay S (1990) Collagen profile in the transplanted heart. Hum Pathol 21: 424–428
77. Rowan RA, Billingham ME (1988) Myocardial innervation in long-term cardiac transplant survivors: a quantitative ultrastructural survey. J Heart Transplant 7: 448–452
78. Kurisu Y, Mastuura Y, Sueda T, Orihashi K, Wada S, Mukai S, Kajihara H, Kato Y (1995) Histologic changes of autonomic nerves following heterotopic cardiac transplantation in rats. Transplant Proc 27: 1565–1567
79. Bahrati S, Billingham ME, Lev M (1992) The conduction system in transplanted hearts. Chest 102: 1182

80. Stovin PGI, Hewitt S (1986)The conduction tissue in the transplanted human heart. J Pathol 149:183
81. Winters GL, Costanzo-Nordin MR (1991) Pathological findings in 2300 consecutive endomyocardial biopsies. Mod Pathol 4: 441–448
82. Pardo-Mindán FJ, Herreros J, Marigil MA, et al. (1986) Myocardial calcification following heart transplantation. J Heart Transplant 5: 332–335
83. Cohnert TR, Kemnitz J (1988) Myocardial calcification after orthotopic heart transplantation. J Heart Transplant 7: 304–308
84. Billingham ME (1994) Pathology and etiology of chronic rejection of the heart. Clin Transplant 8: 289–292
85. Billingham ME (1997) Graft coronary disease: old and new dimensions. Cardiovasc Pathol 6: 95–101
86. Hayry P, Isoniemi H, Yilmaz S, et al. (1993) Chronic allograft rejection. Immunol Rev 134: 33–81
87. Paul LC, Davidoff A, Benediktsson H (1994) Cardiac allograft atherosclerosis in the rat. The effect of histocompatibility factors, cyclosporine, and an angiotensin-converting enzyme inhibitor. Transplantation 57: 1767–1772
88. Mehra MR, Ventura HO, Chambers RB, Ramireddy K, Smart FW, Stapleton DD (1997) The prognostic impact of immunosuppression and cellular rejection on cardiac allograft vasculopathy: time for reappraisal. J Heart Lung Transplant 16: 743–751
89. Libby P, Tanaka H (1994) The pathogenesis of coronary arteriosclerosis ("chronic rejection") in transplanted hearts. Clin Transplant 8: 313–318
90. Baas IO, Offerhaus JA, El-Deiry WS, Wu TC, Hutchins GM, Kasper EK, Baughman KL, Baumgartner WA, Chiou CJ, Hayward GS, Hruban RH (1996) The WAF1-mediated p53 growth-suppressor pathway is intact in the coronary arteries of heart transplant recipients. Hum Pathol 27: 324–329
91. Lemstrom K, Koskinen P, Krogerus L, Daemen M, Bruggeman C, Hayry P (1995) Cytomegalovirus antigen expression, endothelial cell proliferation, and intimal thickening in rat cardiac allografts after cytomegalovirus infection. Circulation 92: 2594–2604
92. Dong C, Wilson JE, Winters GL, McManus BM (1996) Human transplant coronary artery disease: pathological evidence for Fas-mediated apoptotic cytotoxicity in allograft arteriopathy. Lab Invest 74: 921–931
93. White WL, Zhang YL, Shelby J, Trautman MS, Perkins SL, Hammond EH, Shaddy RE (1997) Myocardial apoptosis in a heterotopic murine heart transplantation model for chronic rejection and graft vasculopathy. J Heart Lung Transplant 16: 250–255
94. Radio S, Wood S, Wilson J, Lin H, Winters G, McManus B (1996) Allograft vascular disease: comparison of heart and other grafted organs. Transplant Proc 28: 496–499
95. Ewel CH, Foegh ML (1993) Chronic graft rejection: accelerated transplant arteriosclerosis. Immunol Rev 134: 21–31
96. Pardo-Mindan FJ, Panizo A, Lozano MD, Herreros J, Mejia S (1997) Role of endomyocardial biopsy in the diagnosis of chronic rejection in human heart transplantation. Clin Transplant 11: 426–431
97. Clausell N, Molossi S, Rabinovitch M(1993) Increased interleukin-1b and fibronectin expression are early features of the development of the postcardiac transplant coronary arteriopathy in piglets. Am J Pathol 142: 1772–1786
98. Clausell N, Butany J, Molossi S, Lonn E, Gladstone P, Rabinovitch M, Daly PA (1995) Abnormalities in intramyocardial arteries and lack of correlation with abnormal intracoronary ultrasound or endothelial dysfunction in large epicardial coronary arteries. J Am Coll Cardiol 26: 110–119
99. Boyle JJ, Lawrie G, McPhaden AR, Richens D, Lindop GBM (1995) Arterial lesions associated with medial disorganization and fibrosis in endomyocardial biopsies from human cardiac allografts. Histopathology 27: 439–444
100. Winters GL, Schoen FJ (1997) Graft atherosclerosis-induced myocardial pathology in heart transplant recipients: predictive value of endomyocardial biopsy. J Heart Lung Transplant 16: 985–993
101. Neish AS, Loh E, Schoen FJ (1992) Myocardial changes in cardiac transplant-associated coronary arteriosclerosis: potential for timely diagnosis. J Am Coll Cardiol 19: 586–592

102. Clausell N, Butany J, Gladstone P, Lonn E, Liu P, Cardella C, Feindel C, Daly PA (1996) Myocardial vacuolization, a marker of ischemic injury, in surveillance cardiac biopsies post-transplant: correlations with morphologic vascular disease and endothelial dysfunction. Cardiovasc Pathol 5: 29–37
103. Von Scheidt W, Erdmann E (1991) Dilated angiopathy: a specific subtype of allograft coronary artery disease. J Heart lung Transplant 10:698
104. Weis M, von Scheidt W (1997) Cardiac allograft vasculopathy: a review. Circulation 96: 2069–2077
105. Gao SZ, Hunt SA, Schroeder JS, Alderman E, Hill IR, Stinson EB (1994) Does rapidity of development of transplant coronary artery disease portend a worse prognosis? J Heart Lung Transplant 13: 1119–1124
106. Sharples LD, Caine N, Mullins P, et al. (1991) Risk factor analysis for the major hazards following heart transplantation: rejection, infection, and coronary occlusive disease. Transplantation 52: 244–252
107. Trento A, Griffith BP, Fricker FJ, Kormos RL, Armitage J, Hardesty RL (1989) Lessons learned in pediatric heart transplantation. Ann Thorac Surg 48: 617–622
108. O'Connell JB, Dec GW, Goldenberg IF, Starling RC, Mudge GH, Augustine SM, Costanzo-Nordin MR, Hess ML, Hosepund JD, Icenogle TB (1990) Results of heart transplantation for active lymphocytic myocarditis. J Heart Transplant 9: 351–355
109. Gallo P, Baroldi G, Thiene G, et al. (1993) When and why do heart transplant recipients die? A 7-year experience of 1068 cardiac transplants. Virchows Arch 422: 453–458
110. Kriett JM, Kaye MP (1991)The registry of the International Society for Heart and Lung Transplantation: eighth official report. J Heart Lung Transplant 1991; 10: 491–498
111. Gallo P, Agozzino L, Arbustini E, et al. (1994) Immediate causes of death in short-term surviving heart transplant recipients. Cardiovasc Pathol 3: 173–181
112. Cooper DKC, Lanza RP, Oliver S, et al. (1983)Infectious complications after heart transplantation. Thorax 38: 822–828

Pathology of Pulmonary Transplantation

E. ALVAREZ-FERNÁNDEZ

1 Introduction

Pulmonary transplantation has become a widely accepted procedure, which can be performed as a single-organ procedure, as a block of both lungs or in combination with heart transplantation. The main indications for transplantation are parenchymal disease (idiopathic pulmonary fibrosis, sarcoidosis, eosinophilic granuloma, α-1-antitrypsin deficiency, cystic fibrosis, bronchiolitis obliterans, lymphangiomyomatosis) or vascular disease (primary pulmonary hypertension, the Eisenmenger syndrome). The main causes of death in transplanted patients are infection in the early stages and chronic rejection with bronchiolitis obliterans later [1]. Actuarial survival in the whole group 6 years after transplantation is about 40% [2, 3]; therefore, major pathological findings in transplanted lungs are dominated by the consequences of acute and chronic rejection, and infection. Hyperacute rejection does not have a very specific histopathological profile [3].

2 Initial Response to Implantation

The initial response to implantation develops as a result of a variety of different factors (denervation, liquid overload, deficient lymph drainage, endothelial lesions developing after reperfusion). It appears about 72 h after grafting and consists mainly of an accumulation of oedema fluid in perihilar parenchyma in particular, with minimal leucocyte emigration. It can be difficult to differentiate from early rejection or infection when excessive, but the lack of a significant cellular infiltrate is a strong argument against rejection [4].

Current Topics in Pathology
Volume 92, C. L. Berry (Ed.)

3 Pathology of Rejection

3.1 Acute Rejection

Acute rejection can be observed from 2 days to 3 days after transplantation and is a common phenomenon in the first 6 months after grafting. Most patients have more than one rejection episode within the first 3 months and about half have an episode within the first 3 weeks. Clinically, these patients show decreasing pulmonary function, with progressive dyspnoea, hypoxaemia, leucocytosis, cough and fever. The chest X-ray shows hilar and basal infiltrates. Some patients are clinically asymptomatic and show a normal thoracic X-ray, making the clinical diagnosis of rejection impossible. However, biopsy diagnosis of the early phases of rejection is usually avoided and transbronchial biopsy is used as a diagnostic tool only after the first 3 weeks. However, the histological changes involved are then readily demonstrable by this method in both man and experimental animals [5–9].

Changes begin to develop in the neighbourhood of small venules which become surrounded by an inflammatory infiltrate composed mainly of lymphocytes and some macrophages forming cuffs (Fig. 1). This infiltrate progresses in severity and extent and affects the arterial walls. Then, endothelialitis and some polymorphs, eosinophils and lymphocytes can be seen. As rejection progresses, the infiltrate tends to become denser and to affect not only the perivenular and periarterial spaces, but also the wall of small airways and the pleura. Endothelialitis is more evident. This coincides with an increasing number of transformed lymphocytes in the infiltrate and a higher number of eosinophils and polymorphs. The infiltrate also affects the cartilage-bearing bronchi and the phenomenon is called lymphocytic bronchiolitis and bronchitis, occurring in about 30% of mild acute rejections and in about 60% of moderately severe acute rejection episodes. As the intensity of rejection increases, the cellular infiltrate tends to affect the alveolar septae of the surrounding alveoli and lining pneumocytes show necrosis. Hyaline membranes and intraluminal oedema fluid appear (Fig. 2). In more severe cases, areas of necrosis and haemorrhage develop with vascular thrombosis affecting veins and arteries. These findings may also be seen with infection [1, 8].

Small airways may also show widespread necrosis of epithelium with ulceration and invasion of the lumen by plugs of granulation tissue; a finding observed more frequently in those patients whose rejection progresses to bronchiolitis obliterans than in those who do not. Augmented immunosuppression can have a beneficial role in preventing the evolution of bronchiolitis obliterans [10].

Transbronchial biopsy also gives valuable information about the morphological response to therapy, which is manifest by a decrease in number of cells in the peribronchial and perivascular infiltrate, with persistence of some haemosiderin-laden macrophages as a record of previous haemorrhage [11].

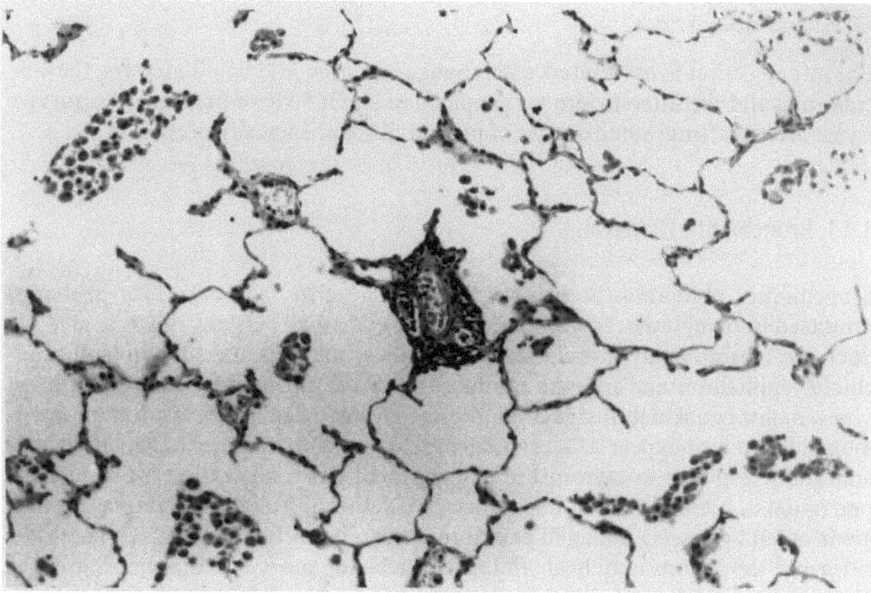

Fig. 1. Mild rejection. Lung parenchyma show small cuffs of lymphoid cells surrounding small venules. H & E, × 100

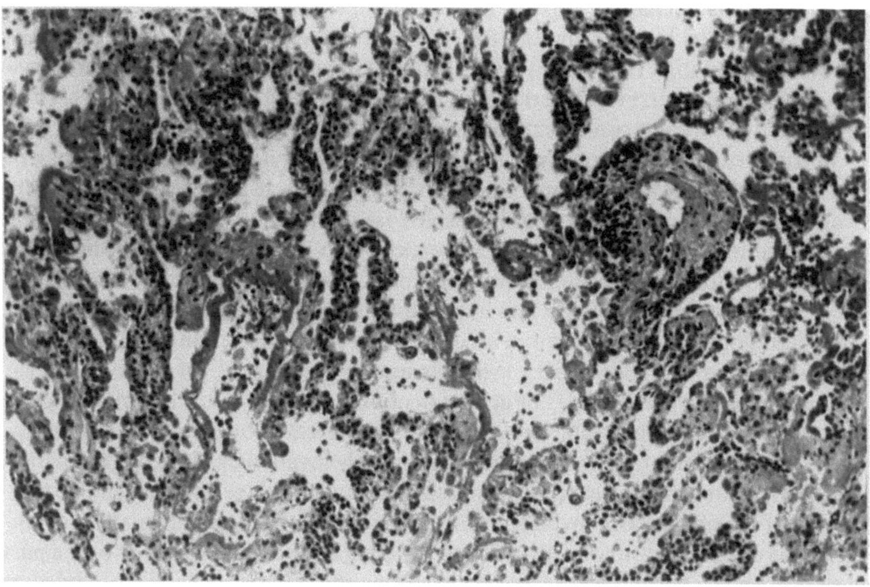

Fig. 2. Severe rejection. In addition to dense perivascular and interstitial cellular infiltrates, alveoli show pneumocyte necrosis and desquamation and prominent hyaline membranes. H & E, × 100

3.2 Chronic Rejection

Chronic rejection is associated with changes in large and small airways, the vascular bed and the interstitium and appears in about 50% of patients who survive 1 year after grafting. Some degree of pleural fibrosis is usually present [4].

3.2.1 Bronchiolitis Obliterans

Bronchiolitis obliterans is a histopathological term that describes a change produced in many ways. Invasion of the lumen of small airways, down to alveolar duct size, by sprouts of granulation tissue follows necrosis and sloughing of bronchiolar epithelium and may be produced by toxic fumes, infections of different types, major bronchial obstruction, diffuse alveolar damage or aspiration pneumonia. It can also appear as a cryptogenic disease with bilateral patchy infiltrates and good response to steroid therapy (bronchiolitis-obliterans type interstitial pneumonia). After the initial injury, necrotic debris, fibrin and inflammatory cells are invaded by vessels, collagen and fibroblasts. The early proliferative phase subsides and the lesions may heal without sequelae or, more commonly, progress to fibrous scarring when the lumen is lost and replaced by a dense collagen scar with some capillary vessels and residual inflammatory cells. In some cases, a concentric band of fibrous tissue develops under the basal lamina, reducing the transverse section of the bronchioli. This justifies the term "constrictive bronchiolitis" preferred by Colby [12]. The epithelium can regenerate or may be replaced by a squamous metaplastic change. Parietal muscle is usually preserved, but in some cases appears distorted by the scar tissue (trichrome and elastic stains help to identify the lost bronchioli in the vicinity of branches of the pulmonary artery). This phase of scarring implies irreversible damage, while in the early proliferative phase some response to treatment can be expected.

In the setting of pulmonary transplantation, obliterative bronchiolitis may be considered the airway component of chronic rejection. It has great clinical significance and is associated with progressive loss of function of small airways and a decrease in pulmonary function. Obstruction to the airflow, cough, increased sputum production and bacterial or fungal infection will follow. In advanced cases, these symptoms and signs do not improve with increase in immunosuppresion. Histopathological changes begin with infiltration of the bronchial wall by activated lymphocytes and is followed by necrosis and desquamation of the bronchiolar epithelium. Repeated episodes of acute rejection infection [4, 13, 15] and ischemia may also play a complementary role in the development of bronchiolitis obliterans. Areas of ulceration of the mucosa are subsequently invaded by plugs of granulation tissue, with a matrix of basophilic ground substance, collagen and a cellular population of fibroblasts and myofibroblasts [16] (Fig. 3). The lesion can affect extensive areas of the small airway or a smaller part in a segmental fashion, or only a part of them (Fig. 4). The clinical severity will rely on the number of totally occluded or severely stenotic bronchioli (Figs. 5–7).

The disease affects both lungs in a patchy fashion and is usually extensive, as in most cases of diffuse damage to small airways associated with functional alterations.

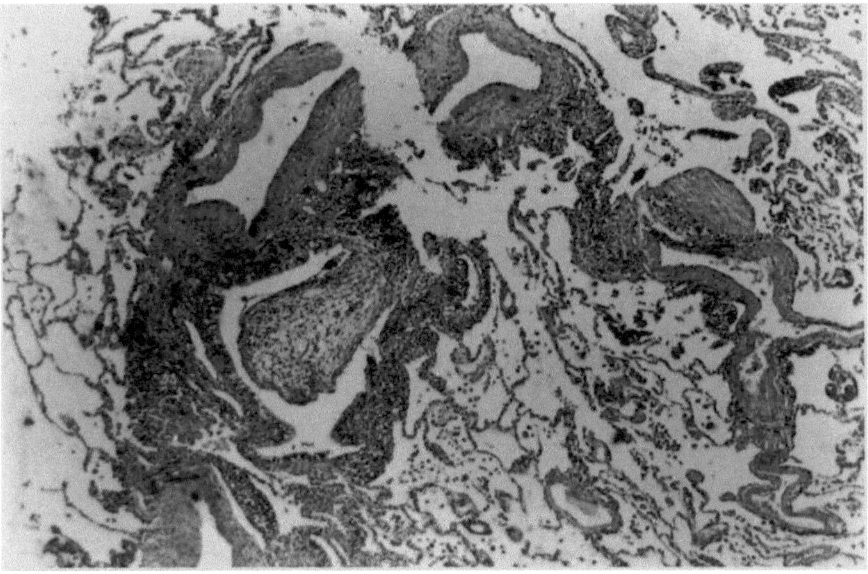

Fig. 3. Active bronchiolitis obliterans in transbronchial biopsy. Granulation tissue protrudes into the lumen of a small bronchiole. H & E, × 40

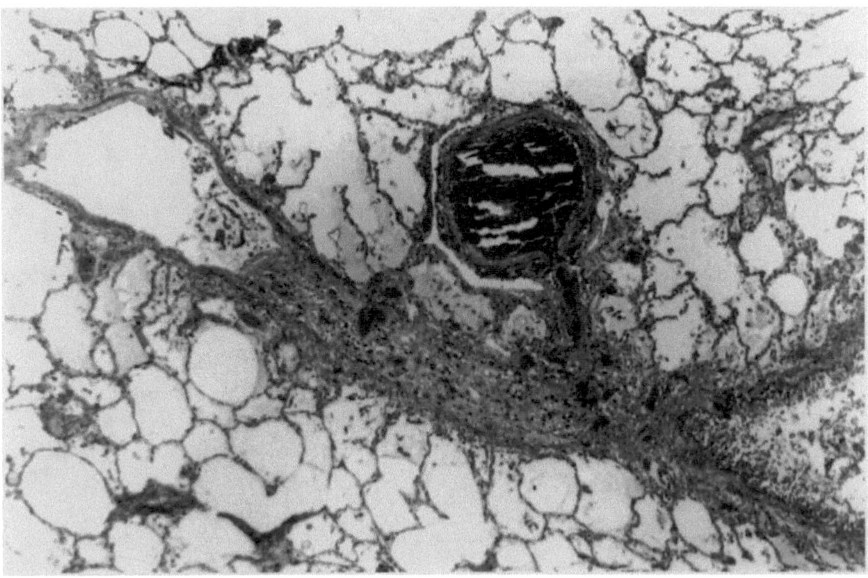

Fig. 4. Chronic rejection. Histological section of an autopsy specimen showing the complete loss of the lumen which seems to be occluded in a segmental fashion. H & E, × 40

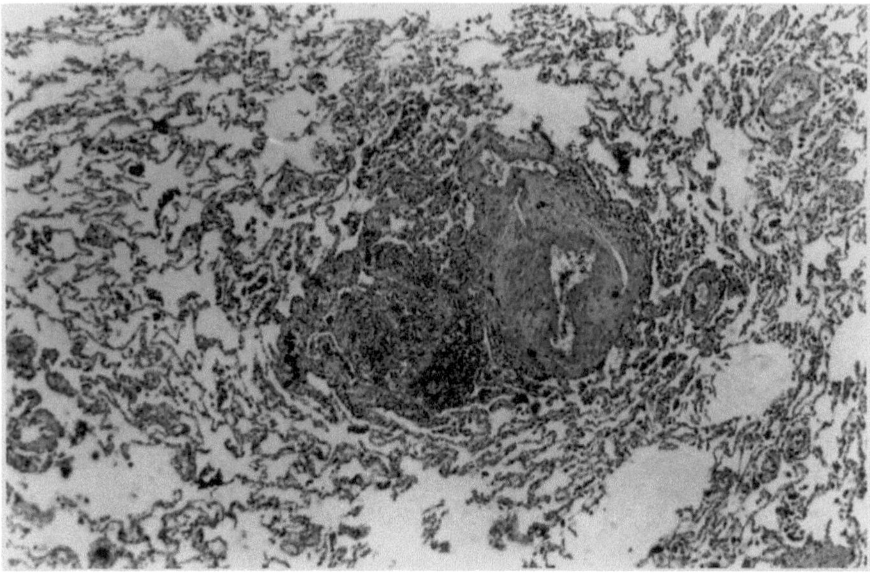

Fig. 5. Low power micrograph to show the complete disappearance of bronchiolar lumen in chronic rejection. Adjacent parenchyma shows a rather unremarkable morphology. H & E, × 40

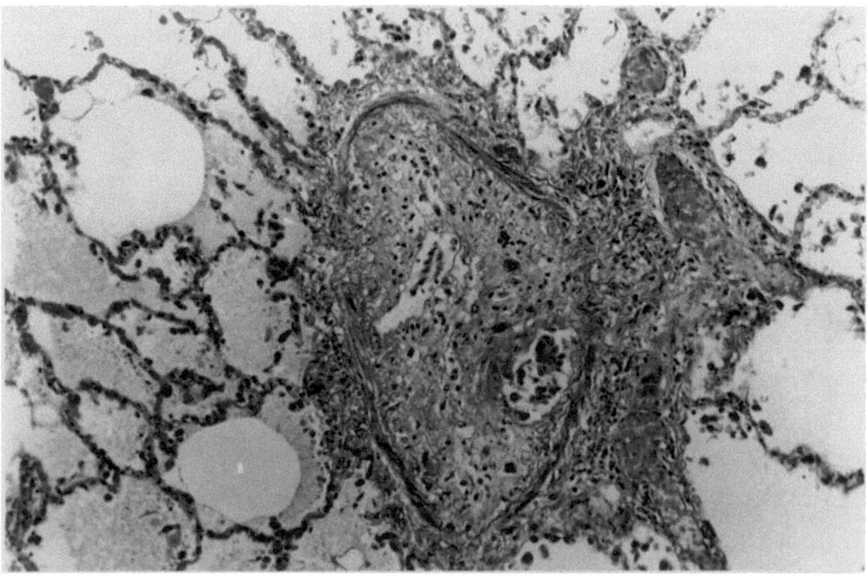

Fig. 6. Chronic rejection. Necropsy specimen to show the substitution of the mucosa and submucosa for dense hyaline collagen with scanty inflammatory cells and two small residual lumens. Bronchiolar muscle persists at the periphery of the scar. H & E, × 100

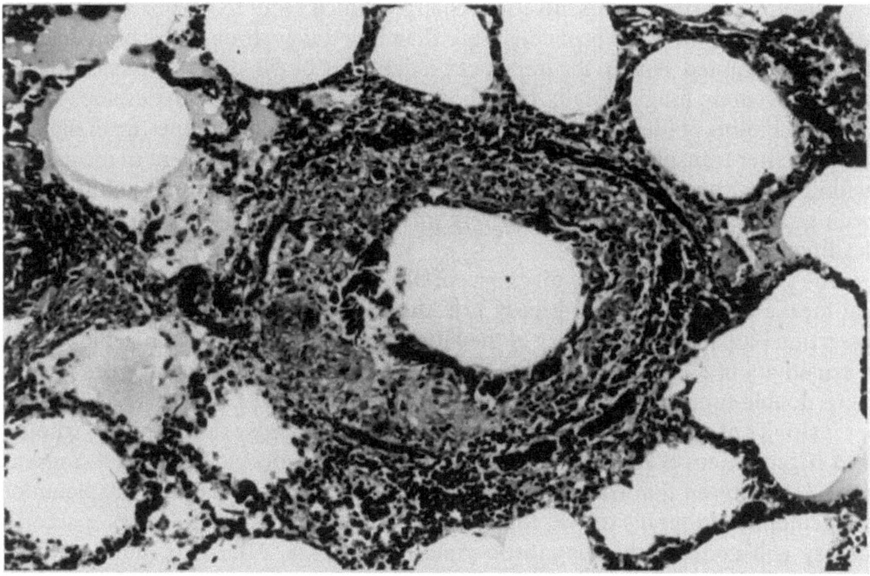

Fig. 7. Constrictive bronchiolitis. Under bronchiolar epithelium, a concentric band of collagen has developed, reducing uniformly the lumen without complete occlusion. H & E, × 100

Recently Abernathy et al. [14], have suggested the existence of two different types of bronchiolitis obliterans. The first one is the "pure form" in which the bronchioli show reduction of their internal diameter due to increase in the amount of collagen in the submucosa or are transformed into solid cords filled with dense scar tissue. This appears late after transplantation and is related to immunological damage to the bronchial epithelium. Adjacent alveoli remain normal. The second form appears early after grafting, showing plugs of granulation tissue in the lumens of affected bronchioli. Adjacent alveoli participate in the inflammatory process. Aetiologically, this change may be related to non-immunological causes, including aspiration or infection. These authors do not exclude the possibility that both morphological forms represent the early and late phases of the same phenomenon.

Bronchiolectasis with mucostasis can be seen proximal to the point of obstruction and as foci of obstructive golden pneumonia, with lipid-laden macropaghes, develop distal to the site of interruption of the lumen. Both are indirect signs of bronchiolar obstruction which may be informative when the point of scarring is not seen in a section of an open biopsy or in autopsy material.

Large cartilaginous bronchi develop bronchiectasis, probably by a combination of causes, including rejection, denervation, ischaemia and recurrent infection [17]. Dilated and inflamed proximal bronchi show squamous metaplasia, cartilage destruction, submucosal lymphoid hyperplasia, chronic inflammation, and atrophy of parietal glands. The loss of cartilage is a major factor in loss of the normal structure and function of the bronchial wall. Yousem et al. [18] have shown how the cartilage plates of the transplanted lung have irregular borders, with perichondral fibrous proliferation and ossification, changes which seem to be closely related to deficient perfusion of the bronchial wall when blood flow through the bronchial artery is interrupted. Finally, the lumen of bronchi may be filled by inspissated mucus that may contain fungi. Distally, bronchocentric granulomatosis can develop.

Bronchiolitis obliterans appears in approximately 30% of patients, from the first months after transplantation to years after grafting [19]. It is evident in most cases within 1 year of transplantation. Two thirds of the patients run a downhill course with progressive obstruction, recurrent infections and poor response to therapy, leading to death.

The gold standard for the diagnosis of bronchiolitis obliterans by morphological means is transbronchial biopsy [7]. Although there is some criticism of this assertion [20]. In the experience of the Pittsburgh group, transbronchial biopsy has a sensitivity of 87% and a specificity of 99%, with no differences between single-lung, double-lung or heart–lung transplantation groups [21]. Scott et al. [15] found a sensitivity of 94% and a specificity of 90% when 18 biopsy samples were studied and Higenbottam et al. [22], a sensitivity of 80% and a specificity of 100%. Yousem et al. (19) showed that transbronchial biopsy may confirm a clinical suspicion of bronchiolitis obliterans in two thirds of patients and, in doubtful cases, a second biopsy will increase the diagnostic yield by 10–20%. After this, thoracoscopic biopsy is indicated if the diagnosis remains doubtful. At least five samples must be examined and three sections of each level stained with Haematoxylin and Eosin (H & E) must be examined. Special stains should be used to exclude infection (Grocot, Zielh, immunohistochemistry) and trichrome stains to evaluate the presence of obliterative change in oblique sections.

3.2.2 Lesions in the Pulmonary Vasculature

Pulmonary arteries and veins may show a variable degree of intimal proliferation. The lumena of elastic and muscular arteries are reduced in diameter by a loose fibrous tissue with spindle-shaped myofibrobastic cells. In some cases, active inflammation with lymphocytes, plasma cells and macrophages is also demonstrated in the intima. Large-size arteries can also show foamy histiocytes in the intima with medial atrophy and reduction in their cross sectional area. All these changes can appear without clinical evidence of significant pulmonary hypertension and correlate well with the number of previous episodes of acute rejection, the presence of bronchiolitis obliterans and similar findings in the coronary arteries in patients with combined heart-lung transplantation [21, 23].

3.3 Grading of Rejection

The clinical importance of the histological diagnosis of rejection and the necessity of having a grading system that permits comparisons of interinstitutional series have encouraged the standardization of criteria in the histopathological diagnosis of lung rejection, resulting in the Working Formulation published in 1990 [5]. Recently, a revision of this formulation was made (Table 1) [6]. In this new report,

Table 1. Working formulation for the classification and grading of pulmonary allograft rejection. (From Yousem et al. 1996 [6])

A. Acute rejection
- Grade 0. None
- Grade 1. Minimal
- Grade 2. Mild
- Grade 3. Moderate
- Grade 4. Severe

B. With/without airway inflammation: lymphocytic bronchitis, bronchiolitis
- BX. Ungradable
- B0. None
- B1. Minimal
- B2. Mild
- B3. Moderate
- B4. Severe

C. Chronic airways rejection: bronchiolitis obliterans
- a. Active
- b. Inactive

D. Chronic vascular rejection: accelerated graft vascular sclerosis

a simplification of the scheme has been achieved, with the emphasis on interstitital infiltrates unrelated to infection in acute rejection and on bronchiolitis obliterans in chronic rejection. The principal tool in the diagnosis of rejection remains transbronchial biopsy. The Working Group recommends obtaining at least five pieces of parenchyma containing bronchioli and at least 100 air sacs to make a proper diagnosis and accurate grading of both acute and chronic rejection. After embedding in paraffin, three levels of the block stained with H&E should be examined. A trichrome stain to evaluate airway and vascular fibrosis and silver stains to demonstrate fungi and/or pneumocystis are mandatory.

The main criterion for the diagnosis of acute rejection is the presence of a mononuclear cell infiltrate around the vessels of peripheral lung parenchyma and in the interstitium. Minimal rejection (grade A1) is characterized by mild perivascular infiltrates which are not evident at scanning magnification, while in mild acute rejection (Grade A2) these infiltrates can be demonstrated easily at low magnification. The cellular infiltration does not extent to adjacent alveolar septa, but endothelialitis and lymphocytic brochiolitis may also be seen. In moderate acute rejection (grade A3) the dense infiltrate affects the perivenular and periarterial spaces and extends to adjacent alveoli. Endothelialitis is very evident and polymorphs and eosinophils are present in the inflammatory infiltrate. Finally, in severe acute rejection (Grade A4), cellular infiltration is associated with necrosis of the alveolar lining pneumocytes, hyaline membranes, haemorrhage and vasculitis.

Vascular infiltrates are therefore the hallmark of acute rejection and the most important criterion in recognition and grading of acute rejection.

Additionally, the presence or absence of inflammation in the small airways is important in predicting the development of bronchiolitis obliterans. Airway inflammation is graded into five different levels: B0 in which no inflammation is present; B1 with minimal infiltration of the bronchial wall by lymphocytes; B2 with a more severe cellular infiltrate, but without significant necrosis or invasion of the epithelium; B3 in which the infiltrate is associated with extensive permeation of the epithelium by lymphocytes with epithelial apoptosis; and B4 in which the most severe lesional complex is present with necrosis of bronchiolar epithelium, ulceration of mucosa and emigration of inflammatory cells, including eosinophils and polymorphs, into the lumen.

The histopathological findings in chronic rejection have been divided into two different components; those associated with the bronchial tree (chronic airway rejection, bronchiolitis obliterans), and those associated with the vascular bed (chronic vascular rejection, accelerated graft vascular sclerosis). Emphasis has been put on the predominance of inflammatory cells and epithelial lesion over the proliferation of dense fibrous collagen tissue in bronchiolitis obliterans. The term active bronchiolitis obliterans thus describes the presence of mononuclear cell mural infiltrates with epithelial damage. Inactive bronchiolitis describes the substitution of the bronchial lumen by dense, almost acellular hyaline collagen. This finding is more significant in establishing the presence of chronic rejection.

4 Infections

Respiratory infections are very common in the lung recipient and are an important cause of morbidity and mortality after transplantation. They are also the major differential diagnosis of rejection as some of the changes associated with rejection can be, at least partially, mimicked by infection [8, 22]. Infection occurs in over 85% of patients, and is facilitated by the absence of the cough reflex, the modification of the mucociliary clearance system, immunosuppression and modification of the bronchial-associated lymphoid tissue. The grafted lung may contribute to infection as a result of infection during the ventilation period before transplantation or from previous infectious disease in the graft. Stewart et al. [24] in emphasizing the importance of a proper study of resected lungs found that unsuspected infections requiring therapy were important. Four cases of active tuberculosis, and two cases aspergillus infection were found in 183 explanted lungs.

Bacterial infection usually appears at two different times after transplantation: during the first 6 weeks and also later when bronchiolitis obliterans and bronchiectasis develop. The principal organisms involved are Pseudomonas, Bacterioides, Serratia and Haemophilus [25]. Mycobacteria, typical and atypical, can also grow in the transplanted lung [26] and this must be excluded when epithelioid granulomas are identified in transbronchial biopsy. This is of particular importance in pulmonary transplantation in sarcoidosis, where recurrence in the graft has been described [27].

Cytomegalovirus (CMV) is the major cause of viral disease in these patients and is associated with an increased incidence of other opportunistic infections. CMV infections may play a role in the development of bronchiolitis obliterans as they enhance the expression of HLA antigens by bronchiolar epithelium. Infected cells show the typical inclusions. The finding of isolated cells with viral inclusions in the absence of inflammatory reaction and the appropriate clinical setting cannot be

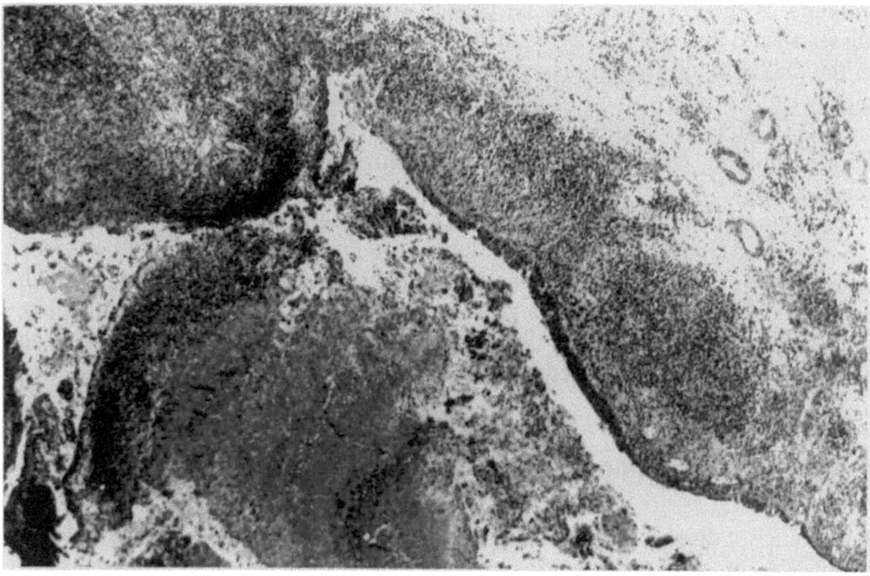

Fig. 8. Bronchiectasis showing dense lymphoid infiltrates and a fungus ball inside the bronchial cavity. H & E, ×40

considered sufficient evidence of CMV pneumonia; indirect signs of CMV infection in the lung have also been suggested, as in the liver (perivascular oedema [28] and neutrophilic abscesses [29]). CMV infection can be diagnosed by serological means, bronchoalveolar lavage and by transbronchial biopsy. The sensitivity of the latter procedure is 83.5% and the specificity 91% [30].

Patients who have undergone transplants may develop herpes tracheobronchitis or pneumonia, which usually appears early after grafting and may be associated with CMV infection. The identification of the characteristic inclusion bodies and immunohistochemistry or molecular biological techniques can be of help in doubtful cases. Other viruses which have been found are paramyxovirus (respiratory syncytial virus and parainfluenza); these are more frequently seen in infantile transplantation than in adults [31].

Aspergillosis is the major fungal disease. Some groups have suggested that its incidence is higher in pulmonary transplantation, when compared with other transplantation groups [32], and have related this to the development of bronchiolitis obliterans. Aspergillus infection occurs in about 15% of transplanted patients and seems to correlate with CMV infection [32]. Different anatomoclinical forms of aspergillosis include superficial non-invasive forms of tracheobronchitis, obstructive tracheobronchitis (with masses of fungi), intrabronchiectatic aspergillomas (Fig. 8) and bronchocentric granulomatosis [33] as well as deep locally invasive pulmonary aspergillosis [27]. Ulcerative invasive aspergillus tracheobronchitis describes a condition where superficial ulceration covered by pseudomembranes is found [32]. This variant is more common at anastomotic sites. Invasive aspergillosis of the lung parenchyma can affect both the transplanted and normal lung [32].

Some of these variants can be treated successfully with antifungal therapy. Disseminated forms are always fatal. Bronchial biopsy and washings can be crucial in the diagnosis of aspergillosis. The demonstration of hyphae in transbronchial biopsy can be done without special techniques, but necrotic tissue must demonstrated to allow the diagnosis of invasive aspergillosis [32]. Infections by other fungi, including Coccidioides inmitis and invasive candidiasis have been described [25].

Pneumocystis infection has a very high incidence after transplantation, but can be controlled by chemoprophylaxis. Its morphology in biopsies may be typical, with foamy exudate into the alveoli and slight interstitial inflammation in the septae, or may be atypical, with a granulomatous form in which clusters of epithelioid histiocytes form sarcoid-like granulomas in the interstitium, in which the organism can be demonstrated [34].

5 Recurrence of Original Disease in the Transplanted Lung

Recurrence of disease is not as frequent as in kidney or hepatic transplantation. In about 80% of cases of sarcoidosis, epithelioid non-infectious granulomas may develop in the grafted lung, but they are not accompanied by abnormal function tests and may resolve. Other recurrent diseases are giant-cell pneumonia and lymphangiomyomatosis [23]. In this latter case, the proliferating smooth muscle cells can be shown to be donor derived [6]. No certain cases of recurrent histiocytosis X, primary pulmonary hypertension or pulmonary fibrosis have been reported.

6 Post-transplant Lymphoproliferative Disorders

As in other transplant recipients, hyperplastic or neoplastic lymphoproliferative disorders are strongly associated with the presence of Epstein-Barr virus. They appear as cellular proliferation of lymphoid cells, sometimes with a very mature morphology and showing plasmocytoid differentiation, and may grow in an interstitial or nodular fashion. When overt lymphoma is present, it is usually of the non-Hodgkin type, arising most commonly from B cells and less frequently from T or natural killer (NK) cells [35]. These lesions tend to arise in extranodal sites, of which the lung graft is a common location [36]. Histological diagnosis by transbronchial biopsy can be difficult due to the clonal polymorphism of some of these proliferations. The presence of prominent blood vessel wall infiltration with vasculitis and coagulative necrosis can be of help [27].

References

1. Chaparro C, Chamberlain D, Maurer J, de Hoyos A, Winton T, Kesten S (1995) Acute lung injury in lung allografts. J Heart Lung Transplant 14: 267–73
2. The Registry of the International Society for Heart and Lung Transplantation (1997) First official pediatric report. J Heart Lung Transplant 16: 1189–1206
3. The Registry of the International Society for Heart and Lung Transplantation (1997) Fourteenth official report. J Heart Lung Transplant 16: 691–712
4. Yousem SA (1994) Transplantation pathology, In: Mario J. Saldaña (ed) Pathology of the lung. Lippincott, Philadelphia, pp 819–826
5. Yousem SA, Berry G, Brunt E, Chamberlain D, Hruban R, Sibley R, Stewart S, Tazelaar H (1990) A working formulation for the standardization of nomenclature in the diagnosis of heart and lung rejection. J Heart Lung Transplant 9: 593–601
6. Yousem SA, Berry GJ, Cagle PT, Chamberlain D et al. (1996) Revision of the 1990 Working Formulation for the Classification of Pulmonary Allograft Rejection: lung rejection study group. J Heart Lung Transplant 15: 1–15
7. Chamberlain D, Maurer J, Chaparro C, Idolor L (1994) Evaluation of transbronchial lung biopsy specimens in the diagnosis of brochiolitis obliterans after lung transplantation. J Heart Lung Transplant 13: 963–71
8. Sibley RK, Berry GJ, Tazelaar HD, Kraemer MR et al. (1993) The role of transbronchial biopsies in the management of lung transplant recipients. J Heart Lung Transplant 12: 308–324
9. Tazelaar HD, Nilsson FN, Rinaldi M, Murtaugh P et al. (1993) The sensitivity of transbronchial biopsy for the diagnosis of acute lung rejection. J Thorac Cardiovasc Surg 105: 674–678
10. Yousem SA (1993) Lymphocytic bronchitis-bronchiolitis in lung allograft recipients. Am J Surg Pathol 17: 491–496
11. Clelland CA, Higenbottam TW, Sterwart S, Scott J, et al. (1990) The histological changes in transbronchial biopsy after treatment of acute lung rejection in heart-lung transplants. J Pathol 161: 105–112
12. Colby TV, Lombard C, Yousem SA, Kitaichi M (1991) Atlas of pulmonary surgical pathology. Saunders, Philadelphia
13. Yousem SA, Dauber JA, Keenan R, Paradis IL, Zeevi A, Griffith BP (1991) Does histologic acute rejection in lung allografts predict the development of bronchiolitis obliterans? Transplantation 52: 306–309
14. Abernathy EC, Hruban RH, Baumgartner WA, Reitz BA, Hutchins GM (1991) The two forms of bronchiolitis obliterans in heart-lung transplant recipients. Hum Pathol 22: 1102–1110
15. Scott JP, Sharples L, Mullins P, Aravot DJ, Stewart S, Otulana BA, Higenbottam TW, Wallwork J (1991) Further studies on the natural history of obliterative bronchiolitis following heart-lung transplantation. Transplant Proc 23: 1201–1202
16. Yousem SA, Burke CM, Billingham ME (1985) Pathologic pulmonary alterations in long-term human heart-ling transplantation. Hum Pathol 16: 911–923
17. Yousem SA, Paradis IL, Dauber JA, Zeevi A et al. (1989) Large airway inflammation in heart-lung transplant recipients. Its significance and prognostic implications. Transplantation 49: 655–656
18. Yousem SA, Dauber JH, Griffith BP (1990) Bronchial cartilage alterations in lung transplantation. Chest 98: 1121–1124
19. Yousem SA, Paradis IL, Dauber JH, Griffith BP (1989) Efficacy of transbronchial lung biopsy in the diagnosis of bronchiolitis obliterans in heart-lung transplant recipient. Transplantation 47: 893–895
20. Kramer MR, Stoehr C, Whang JL, Berry GJ, et al. (1993) The diagnosis of obliterative bronchiolitis after heart-lung and lung transplantation: low yield of transbronchial lung biopsy. J Heart Lung Transplant 12: 675–681
21. Tazelaar HD, Yousem SA (1988) The combined heart-lung transplantation. An autopsy study. Hum Pathol 19: 1403–1416
22. Hogenbottam T, Sterwart S, Penketh A, Wallwork J (1989) Transbronchial lung biopsy for the diagnosis of rejection in heart-lung transplant patients. Transplantation 46: 532–539

23. Yousem SA, Paradis IL, Dauber JH, Zeevi, A et al. (1989) Pulmonary arteriosclerosis in long term human heart-lung transplants recipients. Transplantation 47: 564–569
24. Stewart S, McNeil K, Nashef SAM, Wells FC, et al. (1995) Audit of referral and explant diagnoses in lung transplantation; a pathologic study of lungs removed for parenchymal disease. J Heart Lung Transplant 14: 1173–1186
25. Brooks RG, Hofflin JM, Jamieson SW, Remington JS (1985) Infectious complications in heart-lung transplant recipients. Am J Med 79: 412–422
26. Trulock EP, Bolman RM, Genton R (1989) Pulmonary disease caused by Mycobacterium chelonei in a heart-lung transplant recipient with obliterative bronchiolitis. Am Rev Resp Dis 140: 182–184
27. Stewart S (1994) Lung transplant pathology. In: Hammond EH (ed) Solid organ transplantation pathology. Saunders, Philadelphia, pp 131–158
28. Tazelaar HD (1991) Perivascular inflammation in pulmonary infections: implications for the diagnosis of lung rejection. J Heart Lung Transplant 10: 437–441
29. Stewart S, Higenbottam TW, Hutter JA, Penketh TJ, et al. (1988) Histopathology of transplantation biopsies in heart-lung transplantation. Transplant Proc 20: 764–766
30. Pomerance A, Madden B, Burke MM, Yacoub MH (1995) Transbronchial biopsy in heart and lung transplantation: clinicopathologic correlations. J Heart Lung Transplant 14: 761–773
31. Wendt CH, Fox JMK, Hertz MI (1995) Paramyxovirus infection in lung transplant recipients. J Heart Lung Transplant 14: 479–485
32. Yeldandi V, Laghi F, McCabe MA, Larson R, et al. (1995) Aspergillus and lung transplantation. J Heart Lung Transplant 14: 883–890
33. Tazelaar HD, Baird AM, Mill M, et al. (1989) Bronchocentric mycosis occurring in transplant recipients. Chest 96: 92–94
34. Nine JS, Yousem SA, Paradis IL, Keenan R, Griffith BP (1994) Lymphangioleiomyomatosis: Recurrence after lung transplantation. J Heart Lung Transplant 13: 714–719
35. Harris NL, Ferry JA, Swerdlow SH (1997) Postransplat lymphoproliferative disorders. Summary of Society for Hematopathology Workshop. Semin Diagn Pathol 14: 8–14
36. Yousem SA, Radhawa P, Locker J, et al. (1989) Posttransplant lymphoproliferative disorders in heart-lung transplant recipients. Hum Pathol 20: 361–375

Transplantation in the Central Nervous System *

S. LOVE and D. A. HILTON

1 Introduction

Cells and tissues have been transplanted into the central nervous system (CNS) for a wide range of purposes. These include: (1) studies of factors controlling cell division, migration, growth and differentiation in the CNS; (2) the elucidation of mechanisms of disease; and (3) a means of restoring neurological function, both in animal models of human disease and, increasingly, in human patients. Although the first attempts at CNS transplantation date back to the last century [151], most of the significant scientific contributions in this field have been made during the past two decades. This period has also seen considerable technical advances, includ-

* This manuscript was submitted in February 1997.

Current Topics in Pathology
Volume 92, C. L. Berry (Ed.)

ing improvements in the methods for obtaining, purifying and storing donor cells and tissues [5, 119], the introduction of more effective immunosuppressive protocols for overcoming graft rejection [13, 15, 57, 71, 72, 109], and the development of techniques for introducing, into donor cells, genes that express marker proteins, oncogenes, enzymes for neurotransmitter synthesis, and a range of neurotrophic factors. The result has been a rapid increase in the amount of research and the number of publications in this field. In this chapter, after a brief review of the history of CNS transplantation, the burgeoning literature on its diverse applications is summarised in the context of the different biological and disease processes that have been studied or treated using this technique.

2 Historical Perspective

In 1890, Thomson [151] reported experiments in which he had removed tissue from adult cat cerebral cortex and grafted it into the brains of adult dogs. Histological examination of the brain of one of these dogs at 7 weeks revealed good union of the pia. From the published photomicrographs, it is difficult to be certain whether the transplanted brain tissue survived, although Thomson stated in the text that it had "maintained enough vitality to be distinctly recognised as brain tissue". Subsequent studies included the transplantation of pituitary into the brains of dogs [28] and of neonatal spinal ganglion into the brains of adult rats [117]. Brain to brain transplantation of fragments of cerebral cortex between 10-day-old rat siblings, with histological verification of graft survival and vascularisation, was reported a few years later [31]. The first successful transplants of human brain tumours (glioblastomas and meningiomas) into the anterior chamber of the eye of guinea pigs were described in 1945 [56]. Later experiments resulted in successful transplantation of 8 of 11 human glioblastomas, but none of 10 lower grade tumours, into the brains of guinea pigs and mice [57].

The use of embryonic brain tissue for transplantation was reported by Le Gros Clark in 1940 [79]. Fetal cerebral cortex was implanted into the brains of young rabbits, which were sacrificed 4 weeks later. Histological examination revealed that the transplanted tissue had differentiated and, in some cases, developed a laminar organisation resembling normal cortex. It was not until 1976, however, that Lund and Huschka [90] demonstrated clearly that engrafted neural tissue was capable of forming synaptic interconnections with host neurons.

3 Studies of Developmental Biology

3.1 Segmentation and Embryonic Development of the Nervous System

The embryological development of the CNS is critically dependent on the expression of certain transcription factors in a complex and highly restricted pattern. The temporal and spatial sequence of expression of transcription factors have been well defined in the chick embryo. Transplantation has been used to determine how the expression of these transcription factors is regulated in specific regions of the developing chick CNS. An example concerns the regulation of *Hoxa*-2. The expression of *Hoxa*-2 differs in the developing neural tube and neural crest. To test whether the pattern of *Hox* expression is determined by local cues or is intrinsic to the neuroepithelial cells, i.e., it is pre-specified by their site of origin in the neuroepithelium, Prince and Lumsden [113] transplanted pairs of rhombomeres to ectopic sites at the time of rhombomere boundary formation. They observed that the downregulation of *Hoxa*-2 expression in r2-derived neural crest and maintenance of *Hoxa*-2 expression in r4-derived neural crest is intrinsic to the premigratory crest cell population. Thus, *Hoxa*-2 expression was maintained in r4-derived neural crest that had been grafted to the r2 site and lost in r2-derived neural crest that had been grafted to the r4 site. The expression of several other transcription factors is regulated by local cues. Darnell and Schoenwolf [26] showed that transplants of notochord suppress neural-plate expression of *Engrailed-2* (*En-2*) in the quail, suggesting that the notochord is at least partly responsible for the downregulation of *En-2* in the ventral region of the neural tube during normal embryogenesis.

3.2 Axonal Growth

Neurons that have been obtained from fetal central nervous tissue and transplanted into adult recipients, typically mice or rats, are capable of extending axons a considerable distance within the host nervous system. The route taken depends, in part, on the structure of the white matter tracts into which the axons grow. Thus, transplantation of embryonic mouse hippocampal neurons into the dorsal region of adult rat spinal cords leads to rostral and caudal extension of axons in a slender column within the part of the dorsal tracts occupied by the transplant [80]. Other factors that influence the route taken by axons from engrafted neurons include the presence and proximity of denervated or non-innervated "target" neurons in the host tissue [72, 102, 145]. Lund and Hankin [89] examined the course taken by axons that grew from embryonic retinae implanted in various parts of the brain in mutant mice that lacked prior innervation of their visual centres. These authors found that "retinal" axons followed highly anomalous routes, such as through the internal capsule, to reach the target nuclei in the brain stem. This suggests that the normal optic pathway is not necessary for optic outgrowth and targeting of appropriate second-order neurons in the brain during normal development.

Transplants have also been used as a means of promoting the growth of severed axons. Although the capacity of the adult mammalian CNS to sustain axonal regeneration is normally extremely limited, Aguayo and colleagues have shown that CNS axons are capable of regenerating through implanted segments of peripheral nerve and, in some circumstances, forming functioning synapses on other neurons [12, 17, 27, 75, 121]; these studies are considered below, in Sect. 5.9.

3.3 Neuronal Migration and Differentiation

Synaptic integration of CNS transplants depends not only on appropriate neuronal differentiation with axonal and dendritic extension, but in some cases also on migration of the transplanted precursor cells into the appropriate part of the host CNS. In normal development, many neuronal precursors migrate long distances from the sub-ventricular zone (SVZ) to their final destination within the nervous system. This same process of migration and subsequent neuronal differentiation can be shown to occur when adult mouse SVZ cells that carry a neuron-specific marker transgene are transplanted into the lateral ventricle of other mice [86]. Furthermore, the destination of the implanted cells depends on their original location within the SVZ of the donor animal rather than where they are implanted. Betarbet et al. [6] demonstrated that neuronal precursor cells from the anterior SVZ retained their ability to migrate to the olfactory bulb, even when transplanted heterotopically into the rat striatum.

Much remains to be determined about the stimuli that direct neuronal migration and differentiation. Some of the answers may be provided by transplantation studies. Defects in neuronal migration are a characteristic of the mutant mouse *weaver*. However, when granule cell precursors from *weaver* are implanted into the external germinal layer of wild-type mice, the recipient cells are capable of compensating for these defects and induce normal extension of parallel fibres, migration through the molecular and Purkinje cell layers to the internal granule cell layer and formation of dendrites [51]. Neuronal death provides transplanted cells with a strong stimulus for migration and subsequent differentiation. This is demonstrated by studies in which transplanted embryonic neurons can be induced to repopulate regions of adult mouse neocortex, in which apoptosis has been induced by targeted photolysis [136]. This experiment is also one of several that underline the potential of the CNS to direct region-specific differentiation of transplanted neuronal precursors. Two other experiments serve as illustrations. In the first, Shihabuddin et al. [137] transplanted cells from an immortalised neuronal line, RN33B, into adult and neonatal rat hippocampus and cerebral cortex. In the cerebral cortex, the transplanted cells developed the morphological characteristics of pyramidal and stellate cortical neurons and, in the hippocampus, the transplanted cells differentiated to resemble the pyramidal neurons, granule cells and polymorphic neurons that normally populate this region. The second illustration is provided by experiments in which Vicario-Abejon et al. [162] transplanted dissociated cerebellar cells from newborn rats or mice into the dentate gyrus of the developing hippocampus: the transplanted cells developed morphological, immunohistochemical and ultrastructural features of the host dentate neurons.

3.4 Responses to Trophic Factors

The differentiation, growth and survival of neurons at various stages of development depends, in part, on the provision of specific neurotrophic factors, particularly by the tissues that are targets for innervation. Local infusion of appropriate neurotrophic factors can be used in a wide range of experimental contexts to promote axonal growth or to sustain neurons that would otherwise degenerate. A more convenient and sustainable means of achieving the same end is the transplantation of cells that have been genetically engineered to secrete the neurotrophic factors of interest. By implanting such genetically engineered cells into various regions of the CNS, researchers have been able to determine the effects of trophic factors on different cell populations in vivo. Examples include studies in which fibroblasts genetically modified to secrete nerve growth factor (NGF), brain-derived neurotrophic factor (BDNF), neurotrophin-3 (NT-3), or basic fibroblast growth factor (bFGF) have been grafted into the brain or spinal cord [88, 100]. This strategy has also been employed in several of the studies aimed at restoring function after neuronal degeneration, which are described in Sects. 5.1, 5.4 and 5.8. In some of these, neurotrophin-secreting cells have been co-transplanted with fetal neurons as a means of improving graft survival.

3.5 Glial Migration and Differentiation

Studies of migration and differentiation of transplanted glia and glial precursors have been motivated largely by the recognition of their potential for the treatment of demyelinating disease. Most of the cells for these studies have been derived from fetal tissue [2, 163, 164] or from glial precursor cell lines [9, 45, 131, 152, 156]. These are capable of differentiating into either astrocytes or oligodendrocytes. In hypomyelinated or demyelinated CNS tissue, the oligodendrocytes or their precursors are capable of proliferating, migrating long distances and myelinating or remyelinating axons [2, 78, 152, 163, 164]. However, the transplanted cells are probably able to achieve only very limited migration through normally myelinated white matter [44] (see also Sect. 5.6). In general, transplants of early oligodendrocyte progenitor cells, which are most abundant in tissue from fetuses of relatively early gestational age, have greater capacity for myelination than transplants of more mature oligodendrocytes [2, 164]. The capacity of transplanted oligodendrocytes and Schwann cells to form myelin sheaths and restore nerve conduction is discussed in more detail in Sect. 5.6.

3.6 Hypothalamic Function

Transplantation strategies have been used in studies concerned with at least two aspects of hypothalamic function: sexual development and the control of circadian rhythm. For convenience, in this review, studies of hypothalamic development are considered separately from the use of transplants to restore hypothalamic function, but this division is arbitrary and much of the information is of relevance to both types of research.

Sexual maturation and reproductive function depend on the release of the gonadotrophins luteinising hormone (LH) and follicle-stimulating hormone (FSH) from the anterior pituitary. This is controlled, in turn, by gonadotrophin-releasing hormone (GnRH) neurons in the hypothalamus. In mutant hypogonadal (*hpg*) mice that lack these neurons, sexual maturation can be induced by intracerebral implantation of cells from the GT1 immortalised cell line that secretes GnRH [139]. The GT1 cells migrate widely within the CNS and produce axons, some of which project to the pituitary stalk and stimulate gonadotrophin release. The role of catecholamines and neuropeptides in stimulating hypothalamic LH-releasing hormone (LHRH) neurons and initiating sexual maturation was investigated by Gore et al. [55], who transplanted adrenal medulla into the third ventricle of juvenile female rhesus monkeys. The grafts survived for over 30 months and secreted catecholamines and neuropeptide Y. This resulted in premature ovulation, but did not accelerate other aspects of menarche. Saitoh et al. [128] transplanted either the embryonic olfactory placode, which contains LHRH neurons, or cerebellar tissue, which does not, into the hypothalamus of female rhesus monkeys that had been subjected to bilateral radiofrequency lesions in the arcuate nucleus and the median eminence. Ovulatory cycles, that had been lost after the lesioning, were restored by the olfactory placode grafts, but not by the transplants of cerebellar tissue.

Ralph and Lehman [115] conducted a series of elegant experiments that showed the suprachiasmatic nucleus (SCN) of the hypothalamus to be the site of the circadian pacemaker cells in mammals. By transplanting either the nucleus itself [115] or SCN cells that had been maintained in culture [116], these authors were able to restore normal circadian rhythmicity to hamsters whose nuclei had been ablated, or to convert the circadian rhythm of one strain of hamster to that of a different strain. These experiments have been extended by Kaufman and Menaker [74], who showed that donor tissue from the SCN of hamsters up to postnatal day 12 can be used to restore rhythmicity, and by Sollars et al. [142], who were able to restore circadian rhythms by use of SCN heterografts from mouse to rat and vice versa. In both species, the heterografts elaborated numerous nerve fibres that grew into the host hypothalamus and were still evident 7 months after transplantation.

4 Elucidation of Mechanisms of Disease

4.1 Alzheimer's Disease

Aβ peptides, cleavage products of the larger β-amyloid precursor protein (APP), are the principal constituents of the parenchymal and vascular deposits of amyloid in Alzheimer's disease. Included amongst the many studies devoted to elucidating the possible roles of Aβ peptides and APP in the genesis of this disease, are a small number employing transplantation techniques. Neve et al. [101] observed cortical atrophy and hippocampal abnormalities in adult mice that had been implanted intracerebrally, in the neonatal period, with PC12 (rat phaeochromocytoma) cells transfected with recombinant retrovirus expressing the 104 carboxyl-terminal amino acids of APP. The data were interpreted as suggesting that this part of the precursor protein can cause neurodegeneration, at least when present in excess. In another study, cultured human neurons that secrete Aβ peptides were implanted into mice and rats but, even after survival times of up to 46 weeks, did not cause the formation of plaques or other Alzheimer's disease-like abnormalities [93].

4.2 Creutzfeldt-Jakob Disease

Weissmann, Aguzzi and colleagues conducted a series of ingenious transplantation experiments that illustrate dramatically the key role of host prion protein (PrP) in enabling the spread of spongiform encephalopathies (such as Creutzfeldt-Jakob disease) within the CNS [11, 39]. The transplant recipients were PrP-knockout

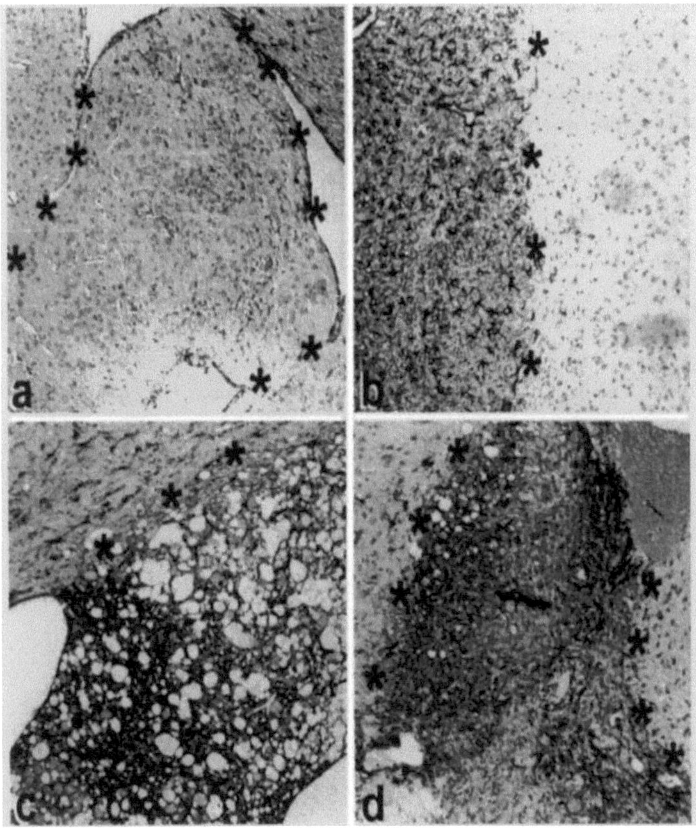

Fig. 1 a–d. Sections through the caudate nucleus of PrP-knockout mice into which PrP^C-expressing tissue has been engrafted. **a** Control mouse, 235 days after intracerebral inoculation with non-infectious brain extract. The PrP^C-expressing graft (outlined by *asterisks*) shows only weak, patchy immunoreactivity for glial fibrillary acidic protein (GFAP), and no spongiform change. **b** Infected mouse, 78 days after inoculation with scrapie extract. The PrP^C-expressing graft (to the *left* of the *asterisks*) shows strong GFAP immunoreactivity. The PrP-knockout host tissue (to the *right* of the *asterisks*) appears normal. **c** Infected mouse, 285 days after inoculation with scrapie extract. Although the PrP-knockout host tissue (towards the *top left* corner) still appears normal, the PrP^C-expressing graft tissue is intensely gliotic (as evidenced by the GFAP immunoreactivity) and shows extensive spongiform vacuolation. **d** Infected mouse, 467 days after inoculation with scrapie extract, most neuronal elements within this graft (between the *asterisks*) have been destroyed and all that remains is densely gliotic tissue. This figure was kindly provided by Professor Adriano Aguzzi, Institut für Neuropathologie der Universität Zürich. (From [11])

mice. These mice are deficient in the normal prion protein, PrP^C, and resistant to spongiform encephalopathy, including scrapie, a form of spongiform encephalopathy that affects sheep and can be transmitted, experimentally, to mice and other animals. Brandner et al. (1996) grafted neural tissue overexpressing PrP^C into the brains of the PrP-knockout mice. Subsequent inoculation of the grafts with scrapie material containing the infectious form of the prion protein (PrP^{Sc}) caused the typical pathological changes of spongiform encephalopathy to develop within the grafts (Fig. 1). However, even though large amounts of graft-derived PrP^{Sc} accumulated in the adjacent host tissue, this remained entirely free of spongiform change (Figs. 2–4), indicating that host PrP is needed for the development of disease. In further experiments, Fischer et al. [39] showed that the susceptibility of PrP-knockout mice to scrapie could be restored by introduction of transgenes encoding either wild-type (normal) PrP or a truncated PrP lacking the 49 amino-

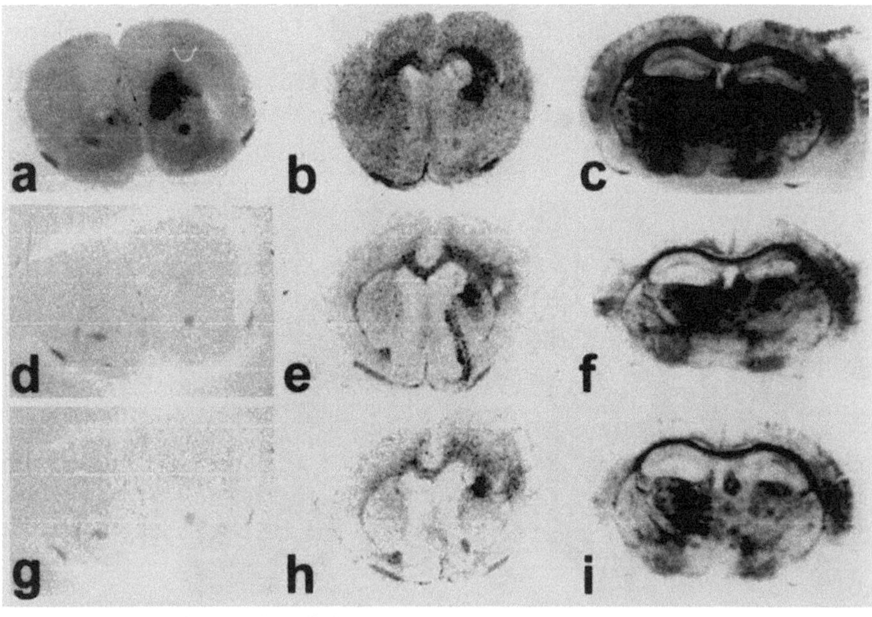

Fig. 2a–i. Accumulation of graft-derived PrP^{Sc} within host brain tissue. Brain histoblots, immunostained for PrP. The first column shows histoblots of PrP^C-engrafted, uninfected brains in PrP-knockout mice, without proteinase K digestion (a), after proteinase K digestion at 20 µg/ml at 55 °C for 8 h (d), or at 50 µg/ml (g). No PrP immunoreactivity is present in the host PrP-knockout tissue. The normal PrP^C in the donor tissue which has been transplanted into the right caudate nucleus is, as expected, susceptible to proteinase-K digestion. The second column shows histoblots of PrP^C-engrafted, scrapie-infected brains in PrP-knockout mice, without (b), with 20 µg/ml (e), and with 50 µg/ml (h) proteinase-K digestion. Proteinase K-resistant PrP expression (indicating the presence of PrP^{Sc}) is most intense within the grafts, but patchy PrP immunoreactivity after proteinase-K digestion is also present within the host brain tissue. The third *column* shows widespread PrP immunoreactivity in histoblots of scrapie-infected brains in wild-type mice, without (c), with 20 µg/ml (f), and with 50 µg/ml (i) proteinase-K digestion. This figure was kindly provided by Professor Adriano Aguzzi, Institut für Neuropathologie der Universität Zürich. (From [11])

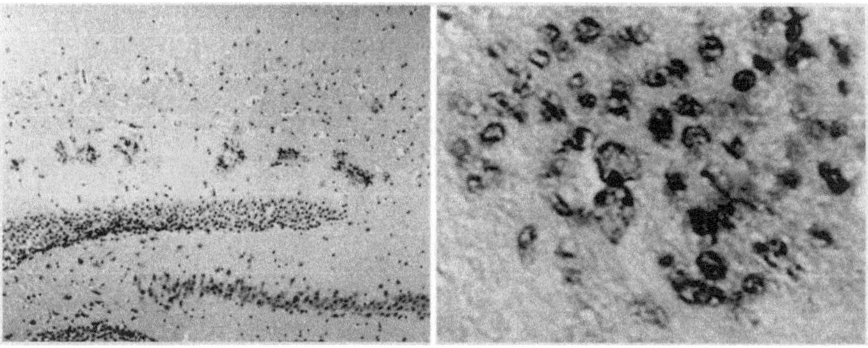

Fig. 3. PrP-immunoreactive deposits in the stratum radiatum and oriens of the hippocampus (*left*) and cerebral cortex (*right*) of scrapie-infected PrP-knockout mice with PrP^C grafts. These deposits are not associated with spongiform degeneration or gliosis of the affected host brain tissue. This figure was kindly provided by Professor Adriano Aguzzi, Institut für Neuropathologie der Universität Zürich. (From [11])

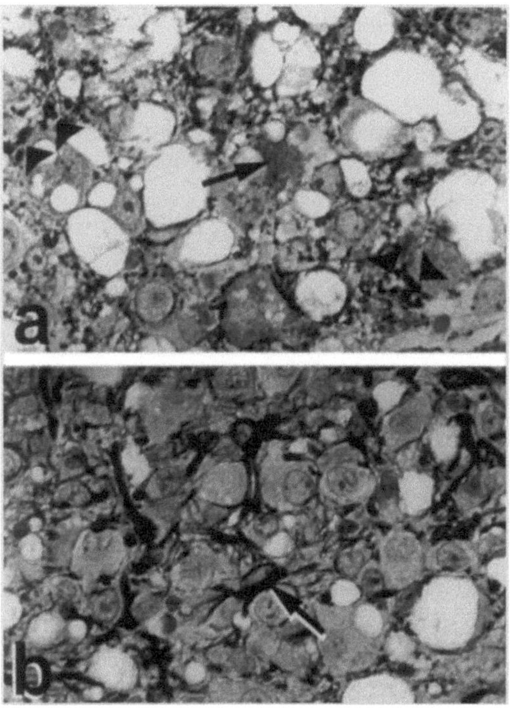

Fig. 4 a, b. Sections through the graft itself, from the mouse illustrated in Fig. 3. **a** The graft shows marked spongiform degeneration, with PrP-immunoreactive deposits (*arrowheads*) and vacuolation of neurons (*arrow*). **b** Immunohistochemistry for glial fibrillary acidic protein (GFAP) reveals striking astrocytosis within the infected graft. This figure was kindly provided by Professor Adriano Aguzzi, Institut für Neuropathologie der Universität Zürich

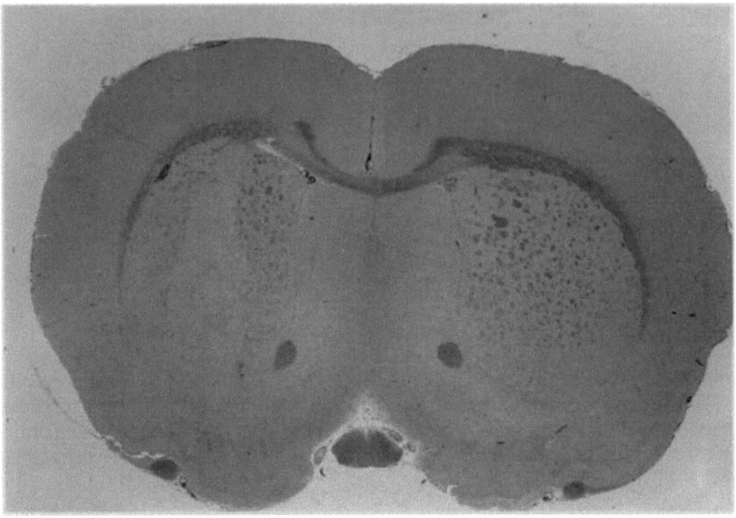

Fig. 5. Coronal section through the brain of a rat into which a suspension of fetal brain cells had been implanted (in the left hemisphere) 8 weeks previously. This figure was kindly provided by Professor Otmar Wiestler, Institut für Neuropathologie der Universitätskliniken Bonn

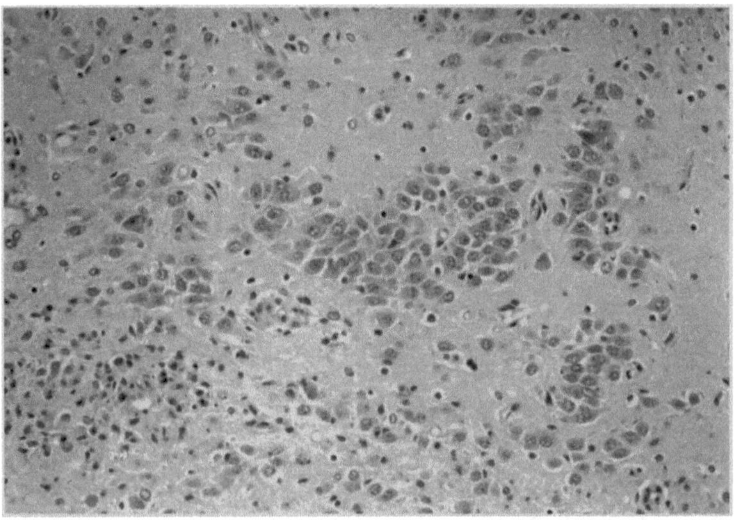

Fig. 6. Histology of the graft shows central nervous system tissue containing neurons and glia. This figure was kindly provided by Professor Otmar Wiestler, Institut für Neuropathologie der Universitätskliniken Bonn

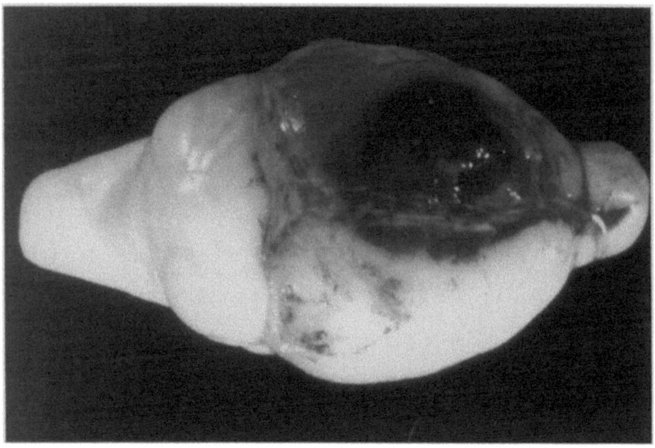

Fig. 7. Haemangioma in a graft expressing the polyoma middle T antigen. This figure was kindly provided by Professor Otmar Wiestler, Institut für Neuropathologie der Universitätskliniken Bonn

proximal amino acids. These amino acids comprise the only part of the protein that is susceptible to protease digestion when it is in an infectious (PrP^{Sc}) form.

4.3 Acquired Immunodeficiency Syndrome (AIDS) Encephalopathy

Xenografts of human neural tissue into immunosuppressed rats and immunodeficient mice have been used to model the encephalopathy caused by human immunodeficiency virus (HIV) infection. Co-engraftment of HIV-1-infected monocytes and human neural tissue causes the grafts to develop many of the neuropathological abnormalities that characterise the AIDS-dementia complex in man [36]. These include neuronal loss, gliosis and the formation of multinucleated giant cells.

4.4 Tumours

Transplantation strategies have been used extensively to investigate the contribution of different oncogenes to the formation of tumours in the CNS [169, 170]. In these studies, fetal brain tissue, usually of rat origin, is infected with replication-defective retroviral vectors that encode the oncogene of interest. It is then implanted into the brain. The grafts give rise to mature CNS tissue (Figs. 5 and 6), within which all cells overexpress the relevant oncogene or, in some experiments, a combination of oncogenes. The development of tumours that are restricted to particular cell types within the grafts is believed to indicate differential susceptibility of those cells to the transforming effects of the oncogene(s) that are overexpressed. Using this approach, Aguzzi et al. [1] showed that the polyoma middle T antigen induces the development of endothelial haemangiomas within weeks (Fig. 7),

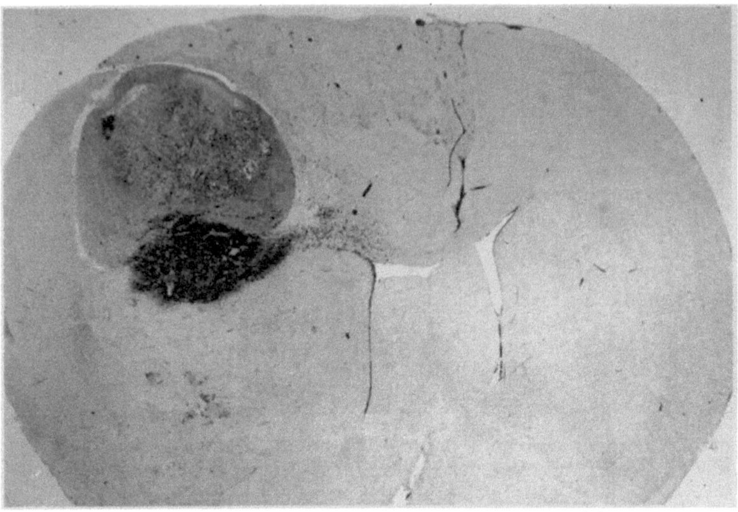

Fig. 8. Astrocytoma in a graft expressing v-*src*. Immunostained for GFAP

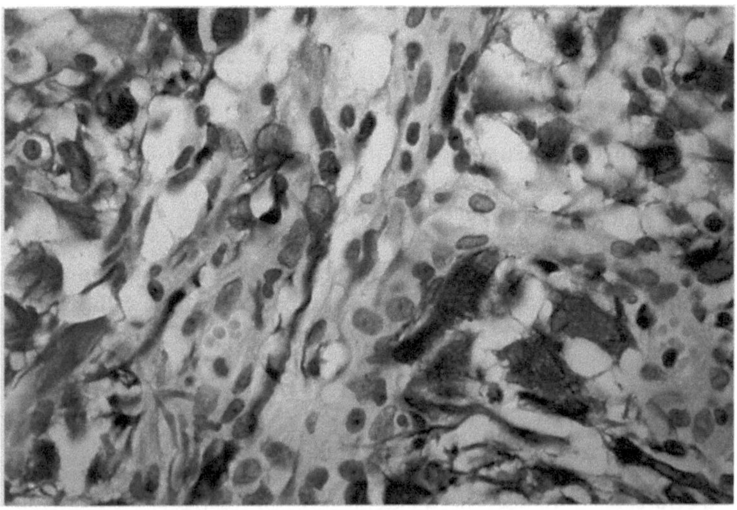

Fig. 9. Higher magnification of a graft expressing v-*src* reveals the astrocytic morphology and cytoplasmic glial fibrillary acidic protein (GFAP) immunoreactivity of the tumour cells. This figure was kindly provided by Professor Otmar Wiestler, Institut für Neuropathologie der Universitätskliniken Bonn

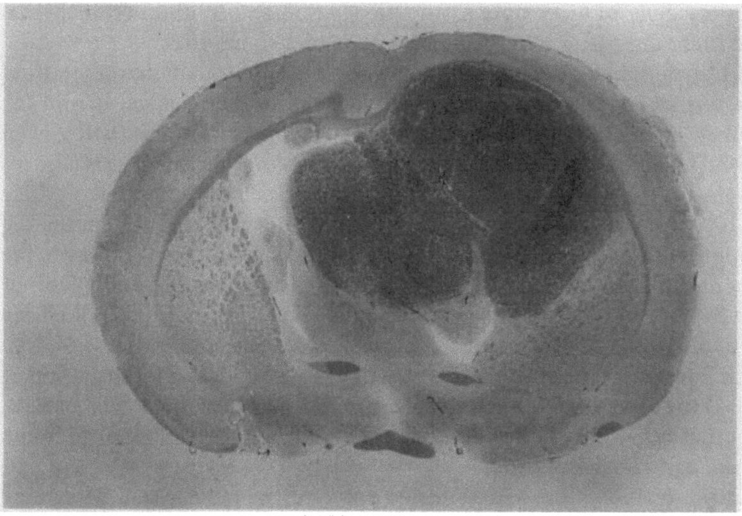

Fig. 10. Primitive neuroectodermal tumour in a graft expressing SV40 virus large T antigen. This figure was kindly provided by Professor Otmar Wiestler, Institut für Neuropathologie der Universitätskliniken Bonn

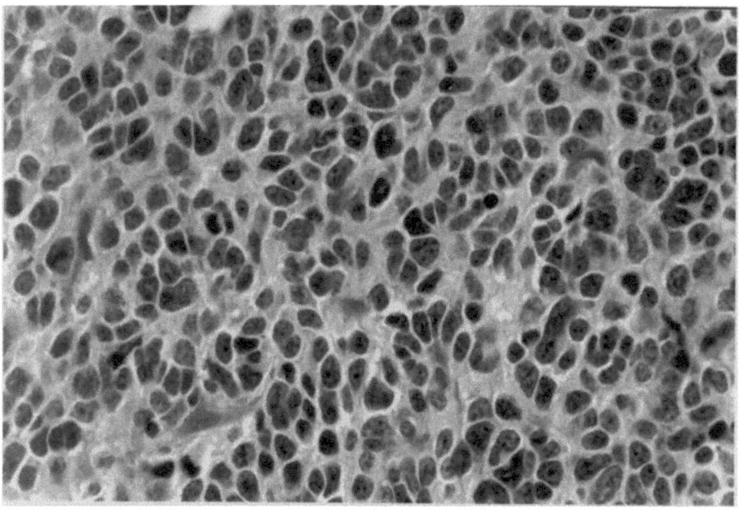

Fig. 11. Higher magnification of a graft expressing SV40 virus large T antigen reveals typical cytological features of a primitive neuroectodermal tumour, including the formation of neuroblastic rosettes. This figure was kindly provided by Professor Otmar Wiestler, Institut für Neuropathologie der Universitätskliniken Bonn

whereas the v-*src* gene induces only astrocytic (Figs. 8 and 9) and mesenchymal tumours, and these only after longer latent periods of 2–6 months.

Retroviral transfer of SV40 virus large T antigen and subsequent neural grafting results in the formation of primitive neuroectodermal tumours (Figs. 10 and 11) that retain their morphological and immunocytochemical characteristics, even after sub-culturing and secondary implantation into further recipients [34]. Introduction of v-Ha-*ras* leads to the development of spindle-cell tumours that produce abundant S-100 protein. Co-expression of v-Ha-*ras* and v-*myc* genes causes the rapid development of highly malignant, polyclonal neoplasms [168]. However, the susceptibility of different cells to the induction of tumours was shown by Radner et al. [114] to depend, in part, on the developmental stage of the animal at the time of exposure to oncogene products: introduction of v-Ha-*ras* and v-*myc* genes into neural cells at later embryonic or early postnatal times induces the formation of endothelial-cell tumours or neuroectodermal tumours showing focal glial or neuronal differentiation, rather than the highly malignant undifferentiated tumours that result from early fetal exposure.

Other transplant studies have investigated some of the factors that determine the grade and invasiveness of gliomas. Deletions involving part or all of chromosome 10 can be demonstrated in at least 50% of glioblastomas. Pershouse and colleagues [111, 146] reintroduced a copy of chromosome 10 into the human glioma cell line U251 and compared the phenotype of the hybrid cells with that of the wild-type U251 cells in intracerebral transplants in nude mice. The reintroduction of chromosome 10 into the human glioblastoma cells completely suppressed their tumorigenic phenotype. In contrast, the introduction of an additional chromosome 2 had no discernible effect on the behaviour of the tumour cells. Edvardsen et al. [33] examined the effects of neural cell-adhesion molecule (NCAM) on the invasion of transplanted BT4Cn rat glioma cells. Transfection of BT4Cn cells with the human transmembrane 140-kDa isoform of NCAM prevented their spreading from the site of transplantation into the adjacent brain parenchyma, suggesting a possible role for NCAM in limiting the invasiveness of gliomas. Gladson et al. [53] showed that the cellular microenvironment influences the expression of the vitronectin, another adhesion molecule thought to be involved in the spread of glial cells, in the U-251MG human astrocytoma line. U-251MG cells implanted subcutaneously in severe combined immunodeficient (SCID) mice did not express detectable vitronectin messenger RNA (mRNA) or protein, whilst intracerebral implants expressed vitronectin at the invading tumour margins.

5 Restoration of Function

Although, as described above, transplants have been widely used to elucidate mechanisms in developmental biology and neurological disease, the driving force behind many CNS transplantation studies has been the goal of preserving or restoring neurological function. Two main strategies have been pursued. The first has been to transplant neurons, glia or other cells to replace those lost as a result of disease. The second, or in some cases supplementary, approach has been to transplant cells that have been genetically engineered to secrete specific trophic factors.

Although much of the research is still experimental, considerable progress has been made in treating animals with models of human disease and, in small numbers, human patients. For further information the reader is referred to several excellent reviews of the potential of transplants for the treatment of neurological disease [18, 30, 47, 49, 62, 82, 135].

5.1 Models of Parkinson's Disease

The animal models that have been most widely used to simulate the loss of nigrostriatal dopaminergic neurons which occurs in Parkinson's disease involve either the injection of 6-hydroxydopamine (6-OHDA) into the striatum or the administration of 1-methyl-4-phenyl-1,2,3,6-tetrahydropyridine (MPTP). Limited observations of relevance to Parkinson's disease have also been made on the mutant mouse *weaver* which, in addition to its other CNS abnormalities, shows depletion of neurons from the substantia nigra and displays motor disturbances, some of which respond to transplantation of dopaminergic neurons [88, 153, 154].

A large number of studies have documented that the motor deficits produced by 6-OHDA or MPTP can be reversed, at least partly, by transplanting dopaminergic cells into the lesioned striatum. The dopaminergic cells used for this purpose by Wolff, Fisher and colleagues [40, 172] were fibroblasts that had been transfected with the transgene for tyrosine hydroxylase, the enzyme which converts tyrosine to L-dopa. The adrenal medulla has also been used as a source of donor dopaminergic cells [25, 42, 46]. However, in most cases, the transplanted cells have been obtained from fetal mesencephalon [97, 102, 138, 141, 147], which is rich in dopaminergic neurons. In animal studies, the number of dopaminergic cells that survive and integrate into the host tissue is greatest if the donor fetuses are of early gestational age [141]. Similar observations have been made after engrafting human fetal tissue from spontaneous abortions into the striatum of rats with 6-OHDA lesions [15, 76]. Rejection of the xenografts can be prevented by administering cyclosporin A [13, 15]. The fetal mesencephalic tissue remains capable of forming viable grafts after it has been stored in liquid nitrogen [119].

Fetal neurons that have been grafted into the striatum or substantia nigra are capable of integrating anatomically and functionally into the local neural circuits [14–16, 21, 97, 102, 118, 138]. However, the mechanism of neurological recovery after transplantation is complex, and involves more than simply the replacement of dopaminergic neurons or reconstitution of striatal circuits [32]. Bankiewicz et al. [4] found that implantation of fetal cerebellum or spinal cord into the caudate nucleus of monkeys with MPTP-induced Parkinsonism ameliorated the neurological deficit, despite the absence of dopaminergic cells in the donor tissues. As discussed earlier, with reference to the studies of Sheen and Macklis [136], Shihabuddin et al. [137] and Vicario-Abejon et al. [162], fetal transplants do, in some situations, acquire the characteristics of local host neurons in the recipient CNS, but this was not found to be the case in the monkeys. The transplants did, however, induce sprouting of dopaminergic fibres from remaining neurons in the ventral striatum and nucleus accumbens. Such sprouting did not occur in control monkeys that had been rendered Parkinsonian with MPTP, but not given any transplants. The results suggest that trophic factor(s) secreted by the transplanted neurons contribute to the recovery of function.

Studies in the early 1980s, by Aguayo and colleagues [12, 17, 27, 75, 99, 121, 122], had shown that CNS axons are capable of growth over relatively long distances through implants of peripheral nerve (see Sect. 5.9). Gage et al. [48] used this approach to promote functional integration between fetal dopaminergic cells that had been placed in the midbrain and the denervated neurons in the striatum of rats with unilateral 6-OHDA lesions. The midbrain grafts were incorporated within the caudal end of a 2-cm to3-cm length of sciatic nerve, the remainder of which was threaded through a burr hole and placed subcutaneously over the skull. After 2 months, the rostral tip of the peripheral nerve graft was cut and inserted through an additional burr hole into the denervated striatum. The authors confirmed that the subsequent functional improvement in some animals was due to growth of axons through the graft, by demonstrating that neurological abnormalities recurred when it was transected subcutaneously.

In early studies, the survival of adrenal autografts was found to be prolonged if NGF was infused into the striatum [148]. In other experiments the adrenal tissue was co-transplanted with sural nerves to provide Schwann cells as a potential sources of NGF [61, 165]. The delivery of trophic factors can be enhanced by implanting genetically engineered cells. Cunningham et al. [25] used a retroviral vector to introduce a mouse β-NGF transgene into rat astrocytes. When the genetically altered astrocytes were co-grafted with adrenal chromaffin cells into the rat striatum, survival of the chromaffin cells was greatly enhanced and they developed an extensive outgrowth of neurites. Chromaffin cells grafted alone or with normal astrocytes did not elaborate neurites. Galpern et al. [50] genetically engineered fibroblasts to produce BDNF. They found that implantation of BDNF-secreting, but not control fibroblasts into the midbrain of MPTP-treated rats increased the levels of dopamine in the substantia nigra, both ipsilateral and contralateral to the graft.

The first attempts at implantation of autologous adrenal medulla in man were carried out in Sweden in 1985 [3], but failed to produce a convincing clinical improvement [83]. Reports from Mexico claimed a dramatic clinical improvement in two young patients with rapidly progressive Parkinson's disease who had received intraventricular adrenal autografts [91]. The improvement in performance was bilateral, although the grafts had been unilateral. This raised the possibility that the effects of the grafts were mediated by release of dopamine into the cerebrospinal fluid (CSF). However, other studies of patients with adrenal autografts have been less impressive [38, 54, 110] and postmortem examinations have not demonstrated graft survival [38, 65, 112]. In one patient, NGF was infused postoperatively into the graft in an attempt to improve its survival; this patient experienced a modest clinical improvement that lasted for nearly 1 year [107].

Lindvall et al. [84] reported a small, but convincing, clinical improvement in two patients with Parkinson's disease who had received striatal implants of tissue from the mesencephalon of 8-week-old foetuses. The transplantation techniques were later improved, leading to more marked clinical improvement in other patients, over at least 3 years [85]. Similar results have been reported by others, the clinical improvement usually starting 1–3 months after surgery [106]. A marked clinical improvement was noted in patients with MPTP-induced Parkinsonism who received striatal grafts [171].

An autopsy study carried out 18 months after a striatal transplant had been performed for Parkinson's disease revealed survival of the grafted tissue, with extensive neuritic outgrowth into the host tissues [77]. Prior to his death, the patient had experienced a sustained clinical improvement, and increased dopamine metabolism in the striatum had been demonstrated by fluorodopa-uptake positron-emission tomography. Despite the absence of immunosuppression during the 12 months prior to death, there was no evidence of graft rejection. Another autopsy study, 4 months after transplantation, also showed survival of the engrafted tissue, with no evidence of cellular rejection [120]. However, Folkerth and Durso [43] demonstrated nodules of cartilage, bone and squamous epithelium, and a lymphocytic infiltrate in the ventricular system of a patient who had seemed to improve neurologically after transplantation and had been receiving cyclosporin for nearly 2 years.

5.2 Models of Huntington's Disease

Striatal degeneration is readily induced in experimental animals by injection of the excitotoxic amino acids, kainic acid, ibotenic acid or quinolinic acid into the caudate nucleus and putamen. Lesioned animals exhibit neuropathological, neurochemical and motor abnormalities resembling those in the basal ganglia of patients with Huntington's disease. A large number of studies have demonstrated that the motor abnormalities can be substantially reversed by engraftment of fetal striatal neurons into the lesioned basal ganglia [58, 67, 69, 104]. Functional recovery has been observed, even after cross-species transplantation [58, 71] although, as applies to other CNS xenografts, rejection occurs unless the recipient is immunosuppressed (usually with cyclosporin A).

Over the course of 5–6 weeks, the transplanted striatal neurons differentiate and form synapses that resemble those in the normal striatum [29]. There is partial restoration of glutamate decarboxylase and choline acetyltransferase activities in the lesioned striatum, and also of glutamate decarboxylase activity in the globus pallidus, which is outside of the graft [68]. Histochemical staining of the mature grafts for acetylcholinesterase, immunohistochemistry for neuropeptides and autoradiographic assessment of receptor binding, generally reveal the types of neuron, receptor and striosomal organisation that are present in intact striatal tissue [70, 71] (Fig. 12), although some organisational, pharmacological and ultrastructural differences persist [29, 104]. To reverse neuronal dysfunction, the transplants have actually to be placed within the caudate nucleus/putamen; Sanberg et al. [129] found implantation into the adjacent lateral ventricle to be without benefit. As is the case in treating animal models of Parkinson's disease, the functional effects are probably not solely attributable to the reconstitution of neuronal circuits by the transplanted neurons. Although the fetal tissue induces vigorous ingrowth of host serotonergic, dopaminergic and cholinergic fibres [174], evidence of reinnervation of the substantia nigra is often scant or lacking [52, 174].

For reasons which are not understood, the transplanted neurons are able to protect the striatum against the noxious effect of further injections of the excitotoxic amino acids, kainic acid or quinolinic acid [158, 159]. This protective effect is

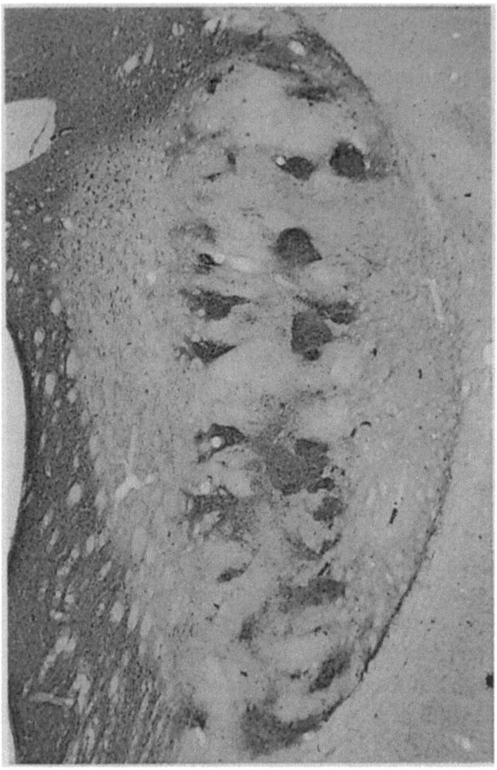

Fig. 12. Neural graft in the rat neostriatum. The striatum had been lesioned with quinolinic acid. The graft was prepared as a cell suspension from E14 rat ganglionic eminence and injected into the striatum 6 h after the lesioning. The graft has been stained immunohistochemically (for DARP-32) to demonstrate the striosomal "patch" regions. This figure was kindly provided by Dr Stephanie Thian, MRC Cambridge Centre for Brain Repair

sustained for at least 30 days after transplantation. In their preliminary report on a patient with Huntington's disease who had been given a unilateral caudate implant of fetal striatal tissue, Madrazo et al. [92] described some clinical improvement at 3 months.

5.3 Cerebellar Degeneration

Two mutant strains of mouse, *weaver* and *pcd* (Purkinje cell degeneration), have been used to study the potential of cerebellar transplants. As noted previously, *weaver* has several defects of neuronal migration and development, including dysplasia of the cerebellar cortex, whilst *pcd* is more akin to degenerative cerebellar ataxia in man. Embryonic wild-type Purkinje cells that have been superficially engrafted into the cerebellum of *pcd* mice migrate radially through the molecular layer, elaborate dendritic trees and induce growth of axons from neurons in the granule cell layer and the formation of appropriate synaptic contacts. This sequence of events closely recapitulates normal development and is complete within 21 days [144]. The fetal grafts not only integrate structurally in a remarkably precise manner, but also partly restore cerebellar function as assessed by a battery of behavioural tests [155, 173].

5.4 Restoring Cognitive Function After Ventral Forebrain or Septohippocampal Lesions (Models of Alzheimer's Disease)

Cognitive defects are readily produced in experimental animals by toxic or surgical lesioning of the nucleus basalis or the forniceal connections between the septal nuclei and hippocampus. These animals can be regarded as models of Alzheimer's disease, only to the extent that the lesions impair cognition and affect some of the parts of the CNS that are also involved in Alzheimer's disease. In the future, mice transgenic for mutant genes that cause the various types of familial Alzheimer's disease may serve as better models for assessing novel therapeutic strategies in that disease. Transplantation studies that use animals with toxic or surgical lesions have, however, provided insight into the potential of transplants for restoring higher neurological function after focal injury.

Loss of cholinergic hippocampal input can be restored by transplanting fetal brain tissue containing septal or nucleus basalis cholinergic neurons into the hippocampus [66, 81, 103, 124, 150]. Cholinergic axons from the transplanted neurons penetrate the hippocampus, release appropriate amounts of acetylcholine and improve performance in radial maze tests and other measures of spatial memory and concentration [66, 81, 103, 124]. Tarricone et al. [150] found the improvement in performance to correlate with the hippocampal cholinergic activity after transplantation. No improvement follows transplantation of non-cholinergic hippocampal tissue into the hippocampus [124]. The results of some studies suggest that optimal restoration of function after forniceal lesions requires replacement of serotonergic as well as cholinergic hippocampal input [73, 123]. Jeltsch et al. [73] compared the effects of grafting cholinergic septal neurons, serotonergic raphe neurons and combinations of these into the hippocampus of rats with lesions of the fimbria and fornix. As would be expected, the grafts of septal neurons increased cholinergic innervation and those of raphe neurons increased serotonergic innervation of the hippocampus, but neither type of graft by itself produced lasting behavioural improvement. However, combined grafts completely normalised performance in the Morris water maze.

Genetically engineered cells, possibly obtained from patients themselves, may prove a more convenient source of acetylcholine than do fetal neurons. Fisher et al. [41] transplanted into the rat hippocampus fibroblasts from a cell line that was genetically modified to express choline acetyltransferase. The fibroblasts continued to produce and release acetylcholine after grafting. Because they cannot integrate anatomically, however, the use of fibroblasts has fairly limited potential. More recently, Gage et al. [47] reported that FGF-2-responsive progenitor cells capable of proliferation and neurogenesis could be isolated from the adult rat hippocampus. On implantation into adult rat brain, these progenitor cells differentiated into mature neurons. This raises the possibility that autologous donor cells capable of genetic modification and functional neuronal integration may be obtainable from the adult human CNS.

The findings of Patel et al. [108] raise the possibility that hippocampal allografts may have deleterious long-term effects on the host tissue. They observed degeneration of neurons in the CA1 field, and abnormal perikaryal and axonal accumulations of phosphorylated neurofilaments in the host hippocampus 12 months after transplantation, despite good graft survival. These abnormalities were significantly more pronounced than in non-grafted, lesion-only rats.

Surgical or toxic (quisqualic acid) lesions of the nucleus basalis cause loss of cholinergic innervation of the neocortex. This can be restored to some extent by implanting fetal brain tissue, containing cholinergic neurons, into the cerebral cortex [103, 125, 126, 130]. As in the case of hippocampal transplants, the neocortical transplants integrate anatomically and functionally, and partly restore cognitive abilities [103, 125]. Although most studies suggest that the cholinergic reinnervation and functional improvement are interrelated, Welner and Koty [166] reported that improvement in sensorimotor, spatial memory and attention tests could be achieved by transplanting non-cholinergic adrenal chromaffin cells into the cerebral cortex of rats with nucleus basalis lesions.

Another transplantation-based therapeutic approach to cholinergic cell-loss lesions merits mention. Martinez-Serrano et al. [95] introduced a vector containing mouse NGF complementary DNA (cDNA), under the control of a retroviral long terminal repeat (LTR) promoter, into a CNS-derived neural progenitor cell line and implanted the cells into the septum of adult rats with a complete fimbria-forniceal lesion. The transplants prevented over 90% of the expected loss of cholinergic neurons from the septal nuclei.

5.5 Transplantation of Motor Neurons

Research in this field has been relatively limited, although it may well increase now that there are mouse models of familial forms of human motor neuron disease caused by mutations in the superoxide dismutase (SOD-1) gene. Ruiz-Flandes et al. [127] monitored the survival and migration of purified embryonic motor neurons that had been fluorescently labelled with carbocyanine and implanted into adult mouse spinal cord or striatum. The motor neurons survived well and migrated long distances from the sites of implantation, some ending up in grey matter, others in white matter. Vrbova and colleagues [22, 23] have shown that transplants of embryonic motor neurons into the spinal cord survive best in cords that lack motor neurons. The transplanted cells migrate from the graft site into the depleted host anterior horns, but once there seem unable to elaborate axons that extend into the ventral nerve roots. It is possible that implantation of "bridges" of peripheral nerve, as described by Aguayo and colleagues [27, 99, 121] (see also Sect. 5.9), may enable transplanted motor neurons to make synaptic contact with skeletal muscle, although this is unlikely to prove practical for treating widespread loss of anterior horn cells.

5.6 Remyelination

Many studies have demonstrated that transplanted oligodendrocytes are capable of forming myelin sheaths around demyelinated axons in the CNS. Although Schwann cells are normally the myelinating cells of the peripheral nervous system, they too readily remyelinate axons in the CNS. It is, indeed, difficult to prevent Schwann cells from contributing, at least in part, to the remyelination of focal lesions induced by injecting lysolecithin, ethidium bromide or other demyelinating agents into the CNS. In early studies of transplanting oligodendrocytes, Blakemore and co-workers [8, 24] found that the injection into demyelinated rat spinal cord of suspensions of "CNS" cells, fewer than 5% of which were contaminating Schwann cells, produced remyelination that was predominantly Schwann-cell mediated.

The model of combined demyelination and destruction of astrocytes, usually produced by injecting ethidium bromide into spinal white matter, previously exposed to 40 Gy of orthovoltage X-irradiation, has been extensively used to study the remyelination that results from glial cell transplantation. The success and relative amounts of oligodendrocyte and Schwann-cell remyelination depend, in part, on the microenvironment into which the cells are injected. When they introduced cultured Schwann cells into demyelinated, astrocyte-free lesions in the cat spinal cord, Blakemore and Crang [7] observed that the subsequent remyelination was largely perivascular or in the vicinity of astrocytes at the periphery of the lesions, suggesting the need for an extracellular matrix or astrocytic framework for Schwann cell-mediated remyelination. Honmou et al. [63] showed that transplants combining cultured Schwann cells and astrocytes produce adequate remyelination to re-establish conduction of nerve impulses through the original lesion. These authors had transfected the cultured cells with a construct containing a *LacZ*-(β-galactosidase) reporter gene and were able to confirm that the remyelination was due to transplanted Schwann cells, rather than invading host cells.

The myelinating capacity of transplanted oligodendrocytes depends on the extent to which they have already differentiated. In experimental animals, the most extensive remyelination is achieved using early oligodendrocyte progenitor cells that have an $A2B5^+$ immunophenotype [2, 9, 164]. Several studies have demonstrated that CNS remyelination can be achieved by implantation of immortalised oligodendrocyte progenitor cell lines [44, 152, 156] (Fig. 13). The paucity of oligodendrocyte progenitor cells in the adult CNS probably explains the failure to achieve remyelination of X-irradiation- and ethidium bromide-induced lesions in rats by implanting oligodendrocyte preparations derived from adult human white matter, despite the survival of the implanted cells [149].

There are conflicting data concerning the extent to which implanted oligodendrocytes are capable of migrating through normal tissue. For example, Vignais et al. [163] used a fluorescent marker dye to trace the migration of oligodendrocyte precursors through the mouse CNS. They found that the precursor cells were able to migrate to and partly remyelinate a demyelinating lesion in the spinal cord after implantation at distances of up to 6–8 mm away from the lesion. However, Franklin et al. [44] reported that cells of a *LacZ*-transfected $O\text{-}2A^+$ progenitor cell line could not migrate through normal white matter, although they were able to migrate through abnormal, X-irradiated rat spinal cord to reach and remyelinate a region of demyelination. The apparent discrepancies may be due to differences in the host and donor species in these studies, or in the techniques used to inject the progenitor cells.

The myelinating capacity of transplanted glia is not confined to focal demyelinating lesions. Transplantation of oligodendrocyte precursors has also proven successful in achieving myelination in animals with dysmyelinating diseases, in

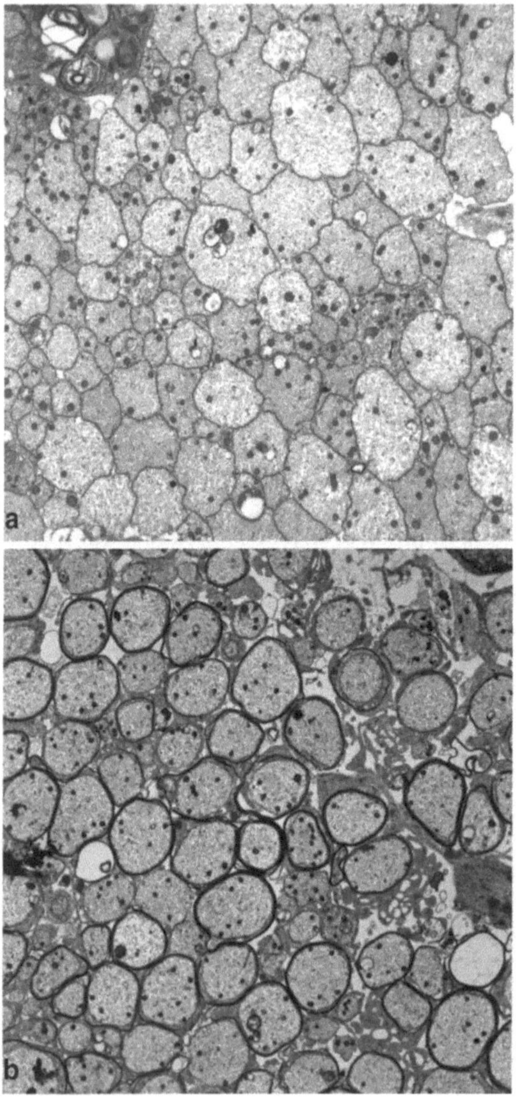

Fig. 13 a, b. **a** Demyelination induced by injecting ethidium bromide into the posterior column of the rat spinal cord after previous exposure of the cord to 40 Gy X-irradiation. **b** Oligodendrocyte-mediated remyelination induced by injection of cells of a *LacZ*-transfected O-2A+ progenitor cell line into the region of demyelination. **c** see p. 203

which myelin sheaths are congenitally deficient. Lachapelle et al. [78] were able to achieve limited myelination in the mutant mouse *shiverer*. Tontsch et al. [152] observed extensive migration of implanted oligodendrocyte precursors and subsequent myelination of the dorsal spinal columns in myelin-deficient rats, and Archer et al. [2], using injections of cells prepared from fetal or neonatal spinal cord, obtained impressive dorsal column myelination in the canine myelin mutant, the *shaking* pup.

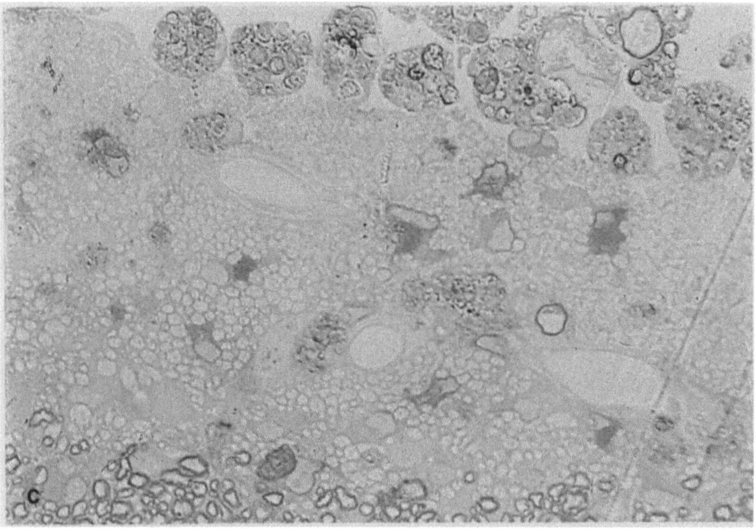

Fig. 13. c The blue X-gal staining of the remyelinating oligodendrocytes confirms that they are exogenous. This figure was kindly provided by Dr R Franklin and Dr W. Blakemore, MRC Cambridge Centre for Brain Repair

5.7 Restoration of Hypothalamic Function

Some aspects of hypothalamic transplantation were described above (see Sect. 3.6). Early work in this field demonstrated that pituitary tissue that had been grafted into the third ventricle of hypophysectomised rats partially restored their growth [60]. Charlton [18] reviewed the use of transplants to restore function in the antidiuretic hormone-deficient Brattleboro rat and the GnRH-deficient *hpg* mouse. A number of experiments have documented that fetal hypothalamic transplants into the ventricles or hypothalamus of Brattleboro rats survive, integrate synaptically and develop appropriate neurovascular relationships [59, 132–134, 167]. Graft survival is improved by use of tissue from relatively early stages of fetal development [10]. No success has been obtained in ameliorating the diabetes insipidus after intraventricular grafting, but implantation of fetal hypothalamic tissue or vasopressin-cell suspensions into the supraoptic region or median eminence has led to functional and biochemical improvement [94, 160].

Charlton et al. [19] showed that the hypogonadism of GnRH-deficient *hpg* mice can be reversed by implanting normal mouse pre-optic area (POA) tissue into the third ventricle. As applies to so many other applications of CNS transplantation, the likelihood of the grafts surviving and functioning is greater if they are obtained from fetal or very early postnatal tissue [20]. More recently, grafts of fetal POA neurons into the POA and of fetal hypothalamic neurons into the ventromedial hypothalamus were shown to reverse some of the masculine sexual behaviour traits of neonatal female rats that had been treated with androgens [157].

5.8 Ageing

As noted previously (see Sect. 5.4), Martinez-Serrano et al. [95] showed that implantation of NGF-secreting neural progenitor cells into the septum of adult rats with a complete fimbria–forniceal lesion prevented loss of cholinergic neurons from the septal nuclei. These authors subsequently examined the effects of grafting NGF-secreting cells into the medial septal nucleus and nucleus basalis magnocellularis of aged rats and, over a 2.5-month period, found significant recovery of cognitive functions [96]. Transplantation has also been found to reverse some age-related changes in circadian function in old hamsters [161]: implantation of fetal SCN tissue, but not cerebellar tissue, into the hypothalamus of these hamsters restored the responsiveness of their circadian clock to the phase-shifting action of triazolam.

5.9 Trauma

Transplant-related strategies for the treatment of trauma can be divided into those aimed at promoting axonal regeneration and those concerned with replacing damaged or lost neurons. The former were extensively investigated by Aguayo and colleagues. They found that CNS axons, incapable of regenerating within central nervous tissue, are able to grow for relatively long distances through "bridges" of peripheral nerve, in which the axons are ensheathed by Schwann cells [27, 121]. When 2-cm to 3-cm lengths of autologous hamster or rat sciatic nerve were implanted between the ocular stump of a transected optic nerve, proximally, and the superior colliculus, distally, retinal ganglion cell axons regenerated through the graft and made synaptic contact with neurons in the superior colliculus [12, 17]. Extracellular recordings of electrical activity in the reinnervated superior colliculi revealed excitatory and inhibitory postsynaptic responses to visual stimulation [75].

Further evidence of the functionality of CNS axons that have grown through peripheral nerve grafts was provided by Munz et al. [99]. They recorded the electrical activity of single regenerated nerve fibres teased from sciatic-nerve grafts that had been implanted into the medulla some months earlier, and observed spontaneous and induced electrical activity, resembling that of normal medullary neurons. A limiting factor in the success of CNS axonal regeneration through peripheral nerve seems to be the distance from the neuronal cell body to the site of injury [122]. Long, descending spinal axons are able to regenerate through peripheral nerve grafts after high or low cervical cord transection, whereas no regeneration occurs after transection at thoracic or lumbar levels.

The other transplant-based approach to restoring function after trauma is the grafting of nerve cells into the sites of injury [37, 64, 140]. Grafts of fetal cortex that were placed in cerebral cavities produced by fluid-percussion brain injury in the rat survived, extended axons into the host tissue, and partly suppressed glial scarring [140]. Improvements in behavioural tests were, however, seen only in animals that were also given local infusions of NGF. Escobar et al. [37] found that implants of fetal brain into lesions of the insular cortex restored the ability of the rats to learn conditioned taste aversion and that the recovery was accelerated if the implants were supplemented with NGF.

5.10 Ischaemic Damage/Infarction

The CA1 field of the hippocampus is particularly vulnerable to the effects of cerebral hypoxia. After experimental transient forebrain ischaemia, the CA1 field can be repopulated with neurons by implantation of fetal hippocampal tissue [98, 105]. The implanted neurons show appropriate differentiation, and some functional connections can be demonstrated, although, to our knowledge, recovery of function has not been reported. Relatively sparse afferent and efferent neuronal connections have also been shown to develop between infarcted rat brain and implants of fetal neocortex [35, 143].

5.11 Treatment of Brain Tumours

As noted above (see Sect. 4.4), brain tumours can be readily transplanted or induced within transplants in experimental animals. Nude (athymic) mice, into which gliomas have been transplanted, are convenient models for testing the in vivo effects of novel anti-tumour agents, and have been extensively used for this purpose. In most such studies, the tumours have been implanted subcutaneously, but intracerebral tumour implants have been used in the evaluation of a wide range of drugs, immunotherapies and treatments involving attenuated viruses. However, detailed consideration of this type of research, in which the brain is used simply as a vehicle for maintaining the growth of implanted tumours, is beyond the scope of this article.

References

1. Aguzzi A, Kleihues P, Heckl K, Wiestler OD (1991) Cell type-specific tumor induction in neural transplants by retrovirus-mediated oncogene transfer. Oncogene 6: 113–118
2. Archer DR, Cuddon PA, Lipsitz D, Duncan ID (1997) Myelination of the canine central nervous system by glial cell transplantation: a model for repair of human myelin disease. Nat Med 3: 54–59
3. Backlund EO, Granberg PO, Hamberger B, et al. (1985) Transplantation of adrenal medullary tissue to striatum in parkinsonism. First clinical trials. J Neurosurg 62: 169–173
4. Bankiewicz KS, Plunkett RJ, Jacobowitz DM, Kopin IJ, Oldfield EH (1991) Fetal non-dopaminergic neural implants in parkinsonian primates. Histochemical and behavioral studies. J Neurosurg 74: 97–104
5. Barker RA, Fricker RA, Abrous DN, Fawcett J, Dunnett SB (1995) A comparative study of preparation techniques for improving the viability of nigral grafts using vital stains, in vitro cultures, and in vivo grafts. Cell Transplant 4: 173–200
6. Betarbet R, Zigova T, Bakay RA, Luskin MB (1996) Migration patterns of neonatal subventricular zone progenitor cells transplanted into the neonatal striatum. Cell Transplant 5: 165–178

7. Blakemore WF, Crang AJ (1985) The use of cultured autologous Schwann cells to remyelinate areas of persistent demyelination in the central nervous system. J Neurol Sci 70: 207–234
8. Blakemore WF, Crang AJ, Patterson RC (1987) Schwann cell remyelination of CNS axons following injection of cultures of CNS cells into areas of persistent demyelination. Neurosci Lett 77: 20–24
9. Blakemore WF, Franklin RJ, Crang AJ (1994) Repair of demyelinated lesions by glial cell transplantation. J Neurol 242[Suppl 1]: S61–63
10. Boer GJ, Gash DM, Dick L, Schluter N (1985) Vasopressin neuron survival in neonatal Brattleboro rats; critical factors in graft development and innervation of the host brain. Neuroscience 15: 1087–1109
11. Brandner S, Isenmann S, Raeber A, Fischer M, Sailer A, Kobayashi Y, Marino S, Weissmann C, Aguzzi A (1996) Normal host prion protein necessary for scrapie induced neurotoxicity. Nature 379: 339–43
12. Bray GM, Villegas-Perez MP, Vidal-Sanz M, Aguayo AJ (1987) The use of peripheral nerve grafts to enhance neuronal survival, promote growth and permit terminal reconnections in the central nervous system of adult rats. J Exp Biol 132: 5–19
13. Brundin P, Nilsson OG, Gage FH, Björklund A (1985) Cyclosporin A increases survival of cross-species intrastriatal grafts of embryonic dopamine-containing neurons. Exp Brain Res 60: 204–208
14. Brundin P, Nilsson OG, Strecker RE, Lindvall O, Astedt B, Björklund A (1986) Behavioural effects of human fetal dopamine neurons grafted in a rat model of Parkinson's disease. Exp Brain Res 65: 235–240
15. Brundin P, Strecker RE, Widner H, Clarke DJ, Nilsson OG, Astedt B, Lindvall O, Björklund A (1988) Human fetal dopamine neurons grafted in a rat model of Parkinson's disease: immunological aspects, spontaneous and drug-induced behaviour, and dopamine release. Exp Brain Res 70: 192–208
16. Campbell K, Wictorin K, Björklund A (1995) Neurotransmitter-related gene expression in intrastriatal striatal transplants–II. Characterization of efferent projecting graft neurons. Neuroscience 64: 35–47
17. Carter DA, Bray GM, Aguayo AJ (1989) Regenerated retinal ganglion cell axons can form well-differentiated synapses in the superior colliculus of adult hamsters. J Neurosci 9: 4042–50
18. Charlton HM (1992) Hypothalamic transplantation. Ciba Found Symp 168: 268–275
19. Charlton HM, Jones AJ, Ward BJ, Detta A, Clayton RN (1987a) Effects of castration or testosterone implants upon pituitary function in hypogonadal mice bearing normal fetal preoptic area grafts. Neuroendocrinology 45: 376–380
20. Charlton HM, Jones AJ, Whitworth D, Gibson MJ, Kokoris G, Zimmerman EA, Silverman AJ (1987b) The effects of the age of intracerebroventricular grafts of normal preoptic area tissue upon pituitary and gonadal function in hypogonadal (HPG) mice. Neuroscience 21: 175–181
21. Clarke DJ, Brundin P, Strecker RE, Nilsson OG, Björklund A, Lindvall O (1988) Human fetal dopamine neurons grafted in a rat model of Parkinson's disease: ultrastructural evidence for synapse formation using tyrosine hydroxylase immunocytochemistry. Exp Brain Res 73: 115–126
22. Clowry G, Sieradzan K, Vrbova G (1991a) Grafts of embryonic tissue into spinal cord: a possible strategy for treating neuromuscular disorders. Neuromuscul Disord 1: 87–92
23. Clowry G, Sieradzan K, Vrbova G (1991b) Transplants of embryonic motoneurones to adult spinal cord: survival and innervation abilities. Trends Neurosci 14: 355–357
24. Crang AJ, Blakemore WF (1989) The effect of the number of oligodendrocytes transplanted into X-irradiated, glial-free lesions on the extent of oligodendrocyte remyelination. Neurosci Lett 103: 269–274
25. Cunningham LA, Hansen JT, Short MP, Bohn MC (1991) The use of genetically altered astrocytes to provide nerve growth factor to adrenal chromaffin cells grafted into the striatum. Brain Res 561: 192–202
26. Darnell DK, Schoenwolf GC (1995) Dorsoventral patterning of the avian mesencephalon/metencephalon: role of the notochord and floor plate in suppressing Engrailed-2. J Neurobiol 26: 62–74

27. David S, Aguayo AJ (1981) Axonal elongation into peripheral nervous system "bridges" after central nervous system injury in adult rats. Science 214: 931–933
28. del Conte G (1907) Einpfanzungen von Embryonalem Gewebe im Gehirn. Beitr Path Anat 42: 193–201
29. DiFiglia M, Schiff L, Deckel AW (1988) Neuronal organization of fetal striatal grafts in kainate- and sham-lesioned rat caudate nucleus: light- and electron-microscopic observations. J Neurosci 8: 1112–1130
30. Duncan ID, Milward EA (1995) Glial cell transplants: experimental therapies of myelin diseases. Brain Pathol 5: 301–310
31. Dunn EJ (1917) Primary and secondary findings in a series of attempts to transplant cerebral cortex in the albino rat. J Comp Neurol 27: 565–582
32. Dunnett SB (1995) Functional repair of striatal systems by neural transplants: evidence for circuit reconstruction. Behav Brain Res 66: 133–142
33. Edvardsen K, Pedersen PH, Bjerkvig R, Hermann GG, Zeuthen J, Laerum OD, Walsh FS, Bock E (1994) Transfection of glioma cells with the neural-cell adhesion molecule NCAM: effect on glioma-cell invasion and growth in vivo. Int J Cancer 58: 116–122
34. Eibl RH, Kleihues P, Jat PS, Wiestler OD (1994) A model for primitive neuroectodermal tumors in transgenic neural transplants harboring the SV40 large T antigen. Am J Pathol 144: 556–564
35. Elsayed MH, Hogan TP, Shaw PL, Castro AJ (1996) Use of fetal cortical grafts in hypoxic-ischemic brain injury in neonatal rats. Exp Neurol 137: 127–141
36. Epstein LG, Cvetkovich TA, Lazar ES, DiLoreto D, Saito Y, James H, del Cerro C, Kaneshima H, McCune JM, Britt WJ, et al. (1994) Human neural xenografts: progress in developing an in-vivo model to study human immunodeficiency virus (HIV) and human cytomegalovirus (HCMV) infection. Adv Neuroimmunol 4: 257–260
37. Escobar ML, Russell RW, Booth RA, Bermudez Rattoni F (1994) Accelerating behavioral recovery after cortical lesions. I. Homotopic implants plus NGF. Behav Neural Biol 61: 73–80
38. Fazzini E, Dwork AJ, Blum C, et al. (1991) Stereotactic implantation of autologous adrenal medulla into caudate nucleus in four patients with parkinsonism. Arch Neurol 48: 813–820
39. Fischer M, Rulicke T, Raeber A, Sailer A, Moser M, Oesch B, Brandner S, Aguzzi A, Weissmann C (1996) Prion protein (PrP) with amino-proximal deletions restoring susceptibility of PrP knockout mice to scrapie. EMBO J 15: 1255–1264
40. Fisher LJ, Jinnah HA, Kale LC, Higgins GA, Gage FH (1991) Survival and function of intrastriatally grafted primary fibroblasts genetically modified to produce L-dopa. Neuron 6: 371–380
41. Fisher LJ, Raymon HK, Gage FH (1993) Cells engineered to produce acetylcholine: therapeutic potential for Alzheimer's disease. Ann N Y Acad Sci 695: 278–284
42. Fitzgerald LR, Glick SD, Schneider AS (1989) Effect of striatal implantation of bovine adrenal chromaffin cells on turning behavior in a rat model of Parkinson's disease. Brain Res 481:373–377
43. Folkerth RD, Durso R (1996) Survival and proliferation of nonneural tissues, with obstruction of cerebral ventricles, in a parkinsonian patient treated with fetal allografts. Neurology 46: 1219–1225
44. Franklin RJ, Bayley SA, Blakemore WF (1996) Transplanted CG4 cells (an oligodendrocyte progenitor cell line) survive, migrate, and contribute to repair of areas of demyelination in X-irradiated and damaged spinal cord but not in normal spinal cord. Exp Neurol 137: 263–276
45. Franklin RJ, Bayley SA, Milner R, Ffrench-Constant C, Blakemore WF (1995) Differentiation of the O-2 A progenitor cell line CG-4 into oligodendrocytes and astrocytes following transplantation into glia deficient areas of CNS white matter. Glia 13: 39–44
46. Freed WJ, Morisha JM, Spoor E, et al. (1981) Transplanted adrenal chromaffin cells in rat brain reduces lesion-induced rotational behaviour. Nature 292: 351–352
47. Gage FH, Ray J, Fisher LJ (1995) Isolation, characterization, and use of stem cells from the CNS. Annu Rev Neurosci 18:159–192

48. Gage FH, Stenevi U, Carlstedt T, Foster G, Björklund A, Aguayo AJ (1985) Anatomical and functional consequences of grafting mesencephalic neurons into a peripheral nerve "bridge" connected to the denervated striatum. Exp Brain Res 60: 584–589
49. Gagnon C, Bedard PJ, Di Paolo T (1993) Grafts in the treatment of Parkinson's disease: animal models. Rev Neurosci 4:17–40
50. Galpern WR, Frim DM, Tatter SB, Altar CA, Beal MF, Isacson O (1996) Cell-mediated delivery of brain-derived neurotrophic factor enhances dopamine levels in an MPP^+ rat model of substantia nigra degeneration. Cell Transplant 5: 225–232
51. Gao WQ, Hatten ME (1993) Neuronal differentiation rescued by implantation of Weaver granule cell precursors into wild-type cerebellar cortex. Science 260: 367–369
52. Giordano M, Hagenmeyer Houser SH, Sanberg PR (1988) Intraparenchymal fetal striatal transplants and recovery in kainic acid lesioned rats. Brain Res 446: 183–188
53. Gladson CL, Wilcox JN, Sanders L, Gillespie GY, Cheresh DA (1995) Cerebral microenvironment influences expression of the vitronectin gene in astrocytic tumors. J Cell Sci 108: 947–956
54. Goetz CG, Olanow CW, Koller WC, et al. (1989) Multicenter study of autologous adrenal medullary transplantation to the corpus striatum in patients with advanced Parkinson's disease. N Engl J Med 320: 337–341
55. Gore AC, Saitoh Y, Terasawa E (1996) Effects of adrenal medulla transplantation into the third ventricle on the onset of puberty in female rhesus monkeys. Exp Neurol 140: 172–183
56. Greene HSN, Arnold H (1945) The homologous and heterologous transplantation of brain and brain tumors. J Neurosurg 2: 315–331
57. Greene HSN (1953) The transplantation of human brain tumors to the brains of laboratory animals. Cancer Res 13: 422–426
58. Hantraye P, Riche D, Maziere M, Isacson O (1992) Intrastriatal transplantation of cross-species fetal striatal cells reduces abnormal movements in a primate model of Huntington disease. Proc Natl Acad Sci U S A 89: 4187–4191
59. Harvey AR, Minson JB, Morris MJ, Chalmers JP (1984) Embryonic hypothalamic tissue transplanted to the IVth ventricle of newborn Brattleboro rats. Neurosci Lett 52: 269–274
60. Halasz B, Pupp L, Uhlarik S, et al. (1963) Growth of hypophysectomised rats bearing pituitary transplants in the hypothalamus. Acta Physiol Acad Sci Hung 23: 287–292
61. Hansen JT, Fiandaca MS, Kordower JH, Notter MFD, Gash DM (1990) Striatal adrenal medulla/sural nerve cografts in hemiparkinsonian monkeys. Prog Brain Res 82: 573–580
62. Hitchcock E (1995) Current trends in neural transplantation. Neurol Res 17: 33–37
63. Honmou O, Felts PA, Waxman SG, Kocsis JD (1996) Restoration of normal conduction properties in demyelinated spinal cord axons in the adult rat by transplantation of exogenous Schwann cells. J Neurosci 16: 3199–3208
64. Hoovler DW, Wrathall JR (1991) Implantation of neuronal suspensions into contusive injury sites in the adult rat spinal cord. Acta Neuropathol (Berl) 81: 303–311
65. Hurtig H, Joyce J, Sladek JR, Trojanowski JQ (1989) Postmortem analysis of adrenal medulla to caudate autograft in a patient with Parkinson's disease. Ann Neurol 25: 607–614
66. Ikegami S, Nihonmatsu I, Kawamura H (1991) Transplantation of ventral forebrain cholinergic neurons to the hippocampus ameliorates impairment of radial-arm maze learning in rats with AF64A treatment. Brain Res 548: 187–195
67. Isacson O, Brundin P, Kelly PA, Gage FH, Björklund A (1984) Functional neuronal replacement by grafted striatal neurones in the ibotenic acid-lesioned rat striatum. Nature 311: 458–460
68. Isacson O, Brundin P, Gage FH, Björklund A (1985) Neural grafting in a rat model of Huntington's disease: progressive neurochemical changes after neostriatal ibotenate lesions and striatal tissue grafting. Neuroscience 16: 799–817
69. Isacson O, Dunnett SB, Björklund A (1986) Graft-induced behavioral recovery in an animal model of Huntington disease. Proc Natl Acad Sci U S A 83: 2728–2732
70. Isacson O, Dawbarn D, Brundin P, Gage FH, Emson PC, Björklund A (1987) Neural grafting in a rat model of Huntington's disease: striosomal-like organization of striatal grafts as revealed by acetylcholinesterase histochemistry, immunocytochemistry and receptor autoradiography. Neuroscience 22: 481–497

71. Isacson O, Riche D, Hantraye P, Sofroniew MV, Maziere M (1989) A primate model of Huntington's disease: cross-species implantation of striatal precursor cells to the excitotoxically lesioned baboon caudate-putamen. Exp Brain Res 75: 213–220
72. Isacson O, Deacon TW, Pakzaban P, Galpern WR, Dinsmore J, Burns LH (1995) Transplanted xenogeneic neural cells in neurodegenerative disease models exhibit remarkable axonal target specificity and distinct growth patterns of glial and axonal fibres. Nat Med 1: 1189–1194
73. Jeltsch H, Cassel JC, Neufang B, Kelche C, Hertting G, Jackisch R, Will B (1994) The effects of intrahippocampal raphe and/or septal grafts in rats with fimbria-fornix lesions depend on the origin of the grafted tissue and the behavioral task used. Neuroscience 63: 19–39
74. Kaufman CM, Menaker M (1993) Effect of transplanting suprachiasmatic nuclei from donors of different ages into completely SCN lesioned hamsters. J Neural Transplant Plast 4: 257–265
75. Keirstead SA, Rasminsky M, Fukuda Y, Carter DA, Aguayo AJ, Vidal-Sanz M (1989) Electrophysiologic responses in hamster superior colliculus evoked by regenerating retinal axons. Science 246: 255–257
76. Kondoh T, Pundt LL, Blount JP, Conrad JA, Low WC (1996) Transplantation of human fetal tissue from spontaneous abortions to a rodent model of Parkinson's disease. Cell Transplant 5: 69–75
77. Kordower JH, Freeman TB, Snow BJ, et al. (1995) Neuropathological evidence of graft survival and striatal reinnervation after the transplantation of fetal mesencephalic tissue in a patient with Parkinson's disease. N Engl J Med 332: 1118–1124
78. Lachapelle F, Duhamel Clerin E, Gansmuller A, Baron Van Evercooren A, Villarroya H, Gumpel M (1994) Transplanted transgenically marked oligodendrocytes survive, migrate and myelinate in the normal mouse brain as they do in the shiverer mouse brain. Eur J Neurosci 6: 814–824
79. Le Gros Clark WE (1940) Neuronal differentiation in implanted fetal cortical tissue. J Neurol Psychiatr 3: 263–284
80. Li Y, Raisman G (1993) Long axon growth from embryonic neurons transplanted into myelinated tracts of the adult rat spinal cord. Brain Res 629: 115–127
81. Li YJ, Simon JR, Low WC (1992) Intrahippocampal grafts of cholinergic-rich striatal tissue ameliorate spatial memory deficits in rats with fornix lesions. Brain Res Bull 29: 147–155
82. Lindvall O (1995) Neural transplantation. Cell Transplant 4: 393–400
83. Lindvall O, Backlund ED, Farde L (1987) Transplantation in Parkinson's disease: two cases of adrenal medullary grafts to the putamen. Ann Neurol 22: 457–468
84. Lindvall O, Rhencrona S, Brundin P, et al. (1989) Human fetal dopamine neurons grafted into the striatum in two patients with severe Parkinson's disease: a detailed account of methodology and a 6-month follow-up. Arch Neurol 46: 615–631
85. Lindvall O, Sawle G, Widner H, et al. (1994) Evidence for long-term survival and function of dopaminergic grafts in progressive Parkinson's disease. Ann Neurol 35: 172–180
86. Lois C, Alvarez-Buylla A (1994) Long-distance neuronal migration in the adult mammalian brain. Science 264: 1145–1148
87. Low WC, Triarhou LC, Kaseda Y, Norton J, Ghetti B (1987) Functional innervation of the striatum by ventral mesencephalic grafts in mice with inherited nigrostriatal dopamine deficiency. Brain Res 435: 315–321
88. Lucidi-Phillipi CA, Gage FH, Shults CW, Jones KR, Reichardt LF, Kang UJ (1995) Brain-derived neurotrophic factor-transduced fibroblasts: production of BDNF and effects of grafting to the adult rat brain. J Comp Neurol 354: 361–376
89. Lund RD, Hankin MH (1995) Pathfinding by retinal ganglion cell axons: transplantation studies in genetically and surgically blind mice. J Comp Neurol 356: 481–489
90. Lund RD, Huschka D (1976) Transplanted neural tissue develops connections with host rat brain. Science 193: 582–584
91. Madrazo I, Drucker-Colin R, Diaz V, Martinez J, Torres C, Becerril JJ (1987) Open microsurgical autograft of adrenal medulla to the right caudate nucleus in two patients with intractable Parkinson's disease. N Engl J Med 316: 831–834
92. Madrazo I, Franco-Bourland RE, Cuevas C, et al. (1991) Fetal neural grafting for the treatment of Huntington's disease – a report of the first case. Soc Neurosci Abstracts 17: 902

93. Mantione JR, Kleppner SR, Miyazono M, Wertkin AM, Lee VM, Trojanowski JQ (1995) Human neurons that constitutively secrete Aβ do not induce Alzheimer's disease pathology following transplantation and long-term survival in the rodent brain. Brain Res 671: 333–337
94. Marciano FF, Gash DM (1986) Structural and functional relationships of grafted vasopressin neurons. Brain Res 370: 338–342
95. Martinez-Serrano A, Lundberg C, Horellou P, Fischer W, Bentlage C, Campbell K, McKay RD, Mallet J, Björklund A (1995) CNS-derived neural progenitor cells for gene transfer of nerve growth factor to the adult rat brain: complete rescue of axotomized cholinergic neurons after transplantation into the septum. J Neurosci 15: 5668–5680
96. Martinez-Serrano A, Fischer W, Soderstrom S, Ebendal T, Bjorklund A (1996) Long-term functional recovery from age-induced spatial memory impairments by nerve growth factor gene transfer to the rat basal forebrain. Proc Natl Acad Sci U S A 93: 6355–6360
97. Moukhles H, Amalric M, Nieoullon A, Daszuta A (1994) Behavioural recovery of rats grafted with dopamine cells after partial striatal dopaminergic depletion in a conditioned reaction-time task. Neuroscience 63: 73–84
98. Mudrick LA, Baimbridge KG (1991) Hippocampal neurons transplanted into ischemically lesioned hippocampus: anatomical assessment of survival, maturation and integration. Exp Brain Res 86: 233–247
99. Munz M, Rasminsky M, Aguayo AJ, Vidal-Sanz M, Devor MG (1985) Functional activity of rat brainstem neurons regenerating axons along peripheral nerve grafts. Brain Res 340: 115–125
100. Nakahara Y, Gage FH, Tuszynski MH (1996) Grafts of fibroblasts genetically modified to secrete NGF, BDNF, NT-3, or basic FGF elicit differential responses in the adult spinal cord. Cell Transplant 5: 191–204
101. Neve RL, Kammesheidt A, Hohmann CF (1992) Brain transplants of cells expressing the carboxyl-terminal fragment of the Alzheimer amyloid protein precursor cause specific neuropathology in vivo. Proc Natl Acad Sci U S A 89: 3448–3452
102. Nikkah G, Cunningham MG, Cenci MA, McKay RD, Björklund A (1995) Dopaminergic microtransplants into the substantia nigra of neonatal rats with bilateral 6-OHDA lesions. I. Evidence for anatomical reconstruction of the nigrostriatal pathway. J Neurosci 15: 3548–3561
103. Nilsson OG, Leanza G, Rosenblad C, Björklund A (1993) Basal forebrain grafts in the hippocampus and neocortex: regulation of acetylcholine release. Ann N Y Acad Sci 695: 267–273
104. Norman AB, Giordano M, Sanberg PR (1989) Fetal striatal tissue grafts into excitotoxin-lesioned striatum: pharmacological and behavioral aspects. Pharmacol Biochem Behav 34: 139–147
105. Nunn J, Hodges H (1994) Cognitive deficits induced by global cerebral ischaemia: relationship to brain damage and reversal by transplants. Behav Brain Res 65: 1–31
106. Olanow CW, Kordower JH, Freeman TB (1996) Fetal nigral transplantation as a therapy for Parkinson's disease. Trends Neurosci 19: 102–109
107. Olson L, Backlund EO, Ebendal T, et al. (1991) Intraputaminal infusion of nerve growth factor to support adrenal medullary autografts in Parkinson's disease: one year follow up of the first clinical trial. Arch Neurol 48: 373–381
108. Patel SN, Kershaw TR, Williams J, Gray JA, Lantos PL, Sinden JD (1995) Neuropathological sequelae of long-term allogeneic and syngeneic neural transplantation into the hippocampus. J Neural Transplant Plast 5: 211–222
109. Pedersen EB, Poulsen FR, Zimmer J, Finsen B (1995) Prevention of mouse-rat brain xenograft rejection by a combination therapy of cyclosporin A, prednisolone and azathioprine. Exp Brain Res 106: 181–186
110. Penn RD, Goetz CG, Tanner CM, et al. (1988) The adrenal medullary transplant operations for Parkinson's disease: clinical observations in five patients. Neurosurgery 22: 999–1004
111. Pershouse MA, Stubblefield E, Hadi A, Killary AM, Yung WK, Steck PA (1993) Analysis of the functional role of chromosome 10 loss in human glioblastomas. Cancer Res 53: 5043–5050
112. Peterson DI, Price ML, Small CS (1989) Autopsy findings in a patient who had an adrenal-to-brain transplant for Parkinson's disease. Neurology 39: 235–238

113. Prince V, Lumsden A (1994) Hoxa-2 expression in normal and transposed rhombomeres: independent regulation in the neural tube and neural crest. Development 120: 911–923
114. Radner H, el Shabrawi Y, Eibl RH, Brustle O, Kenner L, Kleihues P, Wiestler OD (1993) Tumor induction by ras and myc oncogenes in fetal and neonatal brain: modulating effects of developmental stage and retroviral dose. Acta Neuropathol (Berl) 86: 456–465
115. Ralph MR, Lehman MN (1991) Transplantation: a new tool in the analysis of the mammalian hypothalamic circadian pacemaker. Trends Neurosci 14: 362–366
116. Ralph MR, Joyner AL, Lehman MN (1993) Culture and transplantation of the mammalian circadian pacemaker. J Biol Rhythms 8 [Suppl]: S83–S87
117. Ranson WS (1909) Transplantation of the spinal ganglion into the brain. Q Bull Northwest Uni Med School, pp 1–4
118. Redmond DE, Sladek JR Jr, Roth RH, Collier TJ, Elsworth JD, Deutch AY, Haber S (1986) Fetal neuronal grafts in monkeys given methylphenyltetrahydropyridine. Lancet 1: 1125–1127
119. Redmond DE Jr, Naftolin F, Collier TJ, Leranth C, Robbins RJ, Sladek CD, Roth RH, Sladek JR Jr (1988) Cryopreservation, culture, and transplantation of human fetal mesencephalic tissue into monkeys. Science 242: 768–771
120. Redmond DE, Leranth C, Spencer DD, et al. (1990) Fetal neural graft survival. Lancet 336: 820–822
121. Richardson PM, McGuinness UM, Aguayo AJ (1980) Axons from CNS neurons regenerate into PNS grafts. Nature 284: 264–265
122. Richardson PM, Issa VM, Aguayo AJ (1984) Regeneration of long spinal axons in the rat. J Neurocytol 13: 165–182
123. Richter-Levin G, Greenberger V, Segal M (1993) Regional specificity of raphe graft-induced recovery of behavioral functions impaired by combined serotonergic/cholinergic lesions. Exp Neurol 121: 256–260
124. Ridley RM, Gribble S, Clark B, Baker HF, Fine A (1992) Restoration of learning ability in fornix-transected monkeys after fetal basal forebrain but not fetal hippocampal tissue transplantation. Neuroscience 48: 779–792
125. Ridley RM, Baker JA, Baker HF, Maclean CJ (1994) Restoration of cognitive abilities by cholinergic grafts in cortex of monkeys with lesions of the basal nucleus of Meynert. Neuroscience 63: 653–666
126. Rosenblad C, Nilsson OG (1993) Basal forebrain grafts in the rat neocortex restore in vivo acetylcholine release and respond to behavioural activation. Neuroscience 55: 353–362
127. Ruiz-Flandes P, Demierre B, Mattenberger L, Kato AC (1993) Migration of purified embryonic motoneurons grafted into adult mouse CNS. Int J Dev Neurosci 11: 525–533
128. Saitoh Y, Luchansky LL, Claude P, Terasawa E (1995) Transplantation of the fetal olfactory placode restores reproductive cycles in female rhesus monkeys (Mucaca mulatta) bearing lesions in the medial basal hypothalamus. Endocrinology 136: 2760–2769
129. Sanberg PR, Giordano M, Henault MA, Nash DR, Ragozzino ME, Hagenmeyer-Houser SH (1989) Intraparenchymal striatal transplants required for maintenance of behavioral recovery in an animal model of Huntington's disease. J Neural Transplant 1: 23–31
130. Santucci AC, Gluck R, Kanof PD, Haroutunian V (1993) Induction of memory and cortical cholinergic neurochemical recovery with combine fetal transplantation and GM1 treatments in rats with lesions of the NBM. Dementia 4: 273–281
131. Sawamura S, Sawada M, Ito M, Nagatsu T, Nagatsu I, Suzumura A, Shibuya M, Sugita K, Marunouchi T (1995) The bipotential glial progenitor cell line can develop into both oligodendrocytes and astrocytes in the mouse forebrain. Neurosci Lett 188: 1–4
132. Scott DE (1984) Fetal hypothalamic transplants: neuronal and neurovascular interrelationships. Neurosci Lett 51: 93–98
133. Scott DE, Sherman DM (1984) Neuronal and neurovascular integration following transplantation of the fetal hypothalamus into the third cerebral ventricle of adult Brattleboro rats. Neurological transplants: I. Brain Res Bull 12: 453–467
134. Scott DE, Sherman D, Gibbs FP, Paull WK, Gash DM (1984) The neuroanatomical and neurovascular organization of normal fetal hypothalamic explants in the third cerebral ventricle of Brattleboro rats with homozygous diabetes insipidus. Peptides 5 [Suppl 1]: 169–183

135. Shannon KM, Kordower JH (1996) Neural transplantation for Huntington's disease: experimental rationale and recommendations for clinical trials. Cell Transplant 5: 339–352
136. Sheen VL, Macklis JD (1995) Targeted neocortical cell death in adult mice guides migration and differentiation of transplanted embryonic neurons. J Neurosci 15: 8378–8392
137. Shihabuddin LS, Hertz JA, Holets VR, Whittemore SR (1995) The adult CNS retains the potential to direct region-specific differentiation of a transplanted neuronal precursor cell line. J Neurosci 15: 6666–6678
138. Shimizu K, Tsuda N, Okamoto Y, Matsui Y, Miyao Y, Tamura K, Yamada M, Nakatani S, Ikeda T, Mogami H (1988) Transplant-induced recovery from 6-OHDA lesions of the nigrostriatal dopamineneurones in mice. Acta Neurochir Suppl (Wien) 43: 149–153
139. Silverman AJ, Roberts JL, Dong KW, Miller GM, Gibson MJ (1992) Intrahypothalamic injection of a cell line secreting gonadotropin-releasing hormone results in cellular differentiation and reversal of hypogonadism in mutant mice. Proc Natl Acad Sci USA 89: 10668–10672
140. Sinson G, Voddi M, McIntosh TK (1996) Combined fetal neural transplantation and nerve growth factor infusion: effects on neurological outcome following fluid-percussion brain injury in the rat. J Neurosurg 84: 655–662
141. Sladek JR Jr, Elsworth JD, Roth RH, Evans LE, Collier TJ, Cooper SJ, Taylor JR, Redmond DE Jr (1993) Fetal dopamine cell survival after transplantation is dramatically improved at a critical donor gestational age in nonhuman primates. Exp Neurol 122: 16–27
142. Sollars PJ, Kimble DP, Pickard GE (1995) Restoration of circadian behavior by anterior hypothalamic heterografts. J Neurosci 15: 2109–2122
143. Sorensen JC, Grabowski M, Zimmer J, Johansson BB (1996) Fetal neocortical tissue blocks implanted in brain infarcts of adult rats interconnect with the host brain. Exp Neurol 138: 227–235
144. Sotelo C (1993) Cell interactions underlying Purkinje cell replacement by neural grafting in the pcd mutant cerebellum. Can J Neurol Sci 20 [Suppl 3]: S43–S52
145. Stafekhina VS, Bragin AG, Vinogradova OS (1995) Integration of hippocampal suspension grafts with host neocortex. Neuroscience 64: 643–651
146. Steck PA, Ligon AH, Cheong P, Yung WK, Pershouse MA (1995) Two tumor suppressive loci on chromosome 10 involved in human glioblastomas. Genes Chromosomes Cancer 12: 255–261
147. Strecker RE, Miao R, Loring JF (1989) Survival and function of aggregate cultures of rat fetal dopamine neurons grafted in a rat model of Parkinson's disease. Exp Brain Res 76: 315–322
148. Stromberg I, Herrera-Marschitz M, Ungerstedt U, Ebendal T, Olsen L (1985) Chronic implants of chromaffin tissue into the dopamine-denervated striatum. Effects on graft survival, fibre growth and rotational behavior. Exp Brain Res 60: 335–349
149. Targett MP, Sussman J, Scolding N, OLeary MT, Compston DAS, Blakemore WF (1996) Failure to achieve remyelination of demyelinated rat axons following transplantation of glial cells obtained from the adult human brain. Neuropathol Appl Neurobiol 22: 199–206
150. Tarricone BJ, Simon JR, Low WC (1993) Intrahippocampal transplants of septal cholinergic neurons: choline acetyltransferase activity, muscarinic receptor binding, and spatial memory function. Brain Res 632: 41–47
151. Thompson WG (1890) Successful brain grafting. N Y Med J 51: 107
152. Tontsch U, Archer DR, Dubois Dalcq M, Duncan ID (1994) Transplantation of an oligodendrocyte cell line leading to extensive myelination. Proc Natl Acad Sci U S A 91: 11616–11620
153. Triarhou LC, Low WC, Ghetti B (1986) Transplantation of ventral mesencephalic anlagen to hosts with genetic nigrostriatal dopamine deficiency. Proc Natl Acad Sci USA 83:8789–8793
154. Triarhou LC, Norton J, Hingtgen JN (1995) Amelioration of the behavioral phenotype in weaver mutant mice through bilateral intrastriatal grafting of fetal dopamine cells. Exp Brain Res 104: 191–198
155. Triarhou LC, Zhang W, Lee WH (1996) Amelioration of the behavioral phenotype in genetically ataxic mice through bilateral intracerebellar grafting of fetal Purkinje cells. Cell Transplant 5: 269–277

156. Trotter J, Crang AJ, Schachner M, Blakemore WF (1993) Lines of glial precursor cells immortalised with a temperature-sensitive oncogene give rise to astrocytes and oligodendrocytes following transplantation into demyelinated lesions in the central nervous system. Glia 9: 25–40
157. Tsai YF, Chen TJ, Pi WP, Tai MY, Huang RL, Chiueh CC, Peng MT (1995) Effects of fetal brain grafting on adult behavioral masculinization and defeminization in neonatally androgenized female rats. Neurosci Lett 190: 97–100
158. Tulipan N, Huang S, Whetsell WO, Allen GS (1986) Neonatal striatal grafts prevent lethal syndrome produced by bilateral intrastriatal injection of kainic acid. Brain Res 377: 163–167
159. Tulipan N, Luo SQ, Allen GS, Whetsell WO (1988) Striatal grafts provide sustained protection from kainic and quinolinic acid-induced damage. Exp Neurol 102: 325–332
160. UHS, Werner R, Wong D (1987) Correction of genetic diabetes insipidus by adult hypothalamic grafts. Transplantation 43: 485–488
161. Van Reeth O, Zhang Y, Zee PC, Turek FW (1994) Grafting fetal suprachiasmatic nuclei in the hypothalamus of old hamsters restores responsiveness of the circadian clock to a phase shifting stimulus. Brain Res 643: 338–342
162. Vicario-Abejon C, Cunningham MG, McKay RD (1995) Cerebellar precursors transplanted to the neonatal dentate gyrus express features characteristic of hippocampal neurons. J Neurosci 15: 6351–6363
163. Vignais L, Nait-Oumesmar B, Mellouk F, Gout O, Labourdette G, Baron Van Evercooren A, Gumpel M (1993) Transplantation of oligodendrocyte precursors in the adult demyelinated spinal cord: migration and remyelination. Int J Dev Neurosci 11: 603–612
164. Warrington AE, Barbarese E, Pfeiffer SE (1993) Differential myelinogenic capacity of specific developmental stages of the oligodendrocyte lineage upon transplantation into hypomyelinating hosts. J Neurosci Res 34: 1–13
165. Watts RL, Bakay RAE, Herring CJ, et al. (1990) Preliminary report on adrenal medullary grafting and cografting with sural nerve in the treatment of hemiparkinson monkeys. Prog Brain Res 82: 581–591
166. Welner SA, Koty ZC (1993) Amelioration of sensory attention and sensorimotor deficits by chromaffin cell grafts to the cerebral cortex of nucleus basalis magnocellularis lesioned rats. Behav Brain Res 59: 73–81
167. Wiegand SJ, Gash DM (1988) Characteristics of vasculature and neurovascular relations in intraventricular anterior hypothalamic transplants. Brain Res Bull 20: 105–124
168. Wiestler OD, Aguzzi A, Schneemann M, Eibl R, von Deimling A, Kleihues P (1992a) Oncogene complementation in fetal brain transplants. Cancer Res 52: 3760–3767
169. Wiestler OD, Brustle O, Eibl RH, Radner H, Aguzzi A, Kleihues P (1992b) Retrovirus-mediated oncogene transfer into neural transplants. Brain Pathol 2: 47–59
170. Wiestler OD, Brustle O, Eibl RH, Radner H, Aguzzi A, Kleihues P (1994) Oncogene transfer into the brain. Recent Results Cancer Res 135: 55–66
171. Winder H, Tetrud J, Rehncrona S, et al. (1992) Bilateral fetal mesencephalic grafting in two patients with parkinsonism induced by 1-methyl-4-phenyl-1,2,3,6-tetrahydropyridine (MPTP). N Eng J Med 327: 1556–1563
172. Wolff JA, Fisher LJ, Xu L, Jinnah HA, Langlais PJ, Iuvone PM, O'Malley KL, Rosenberg MB, Shimohama S, Friedmann T, et al. (1989) Grafting fibroblasts genetically modified to produce L-dopa in a rat model of Parkinson disease. Proc Natl Acad Sci U S A 86: 9011–9014
173. Zhang W, Lee WH, Triarhou LC (1996) Grafted cerebellar cells in a mouse model of hereditary ataxia express IGF-I system genes and partially restore behavioral function. Nat Med 2: 65–71
174. Zhou FC, Buchwald N (1989) Connectivities of the striatal grafts in adult rat brain: a rich afference and scant striatonigral efference. Brain Res 504: 15–30

Subject Index

Index of Volumes 89–91 Current Topics in Pathology

Volume 91: Thyroid Pathology. Edited by K. W. Schmid and W. Böcker

MIX
Papier aus verantwortungsvollen Quellen
Paper from responsible sources
FSC® C105338

If you have any concerns about our products,
you can contact us on
ProductSafety@springernature.com

In case Publisher is established outside the EU,
the EU authorized representative is:
Springer Nature Customer Service Center GmbH
Europaplatz 3, 69115 Heidelberg, Germany

Printed by Libri Plureos GmbH
in Hamburg, Germany